River East, River West

Iowa's Natural Borders

David Hamilto
John T Price
Lisa Knopp
Robert Wolf
Gary Holthaus
Ethan Hirsh

Ice Cube Press
Harvest Book
North Liberty, Iowa

River East, River West: Iowa's Natural Borders

ISBN 978-1-888160-24-6 (1-888160-24-1)

Library of Congress Control Number: 2006926993

Ice Cube Press (est. 1993)
205 N Front Street
North Liberty, Iowa 52317-9302
www.icecubepress.com

Manufactured in the United States of America on recycled paper.

♾ The paper used in this publication meets the minimum requirements of the American National Standard for Information Sciences—Permanence of Paper for Printed Library Materials, ANSI Z39.48-1992

"Why Geese Don't Winter in Paradise" was previously published in *Isotope: A Journal of Literary Nature and Science Writing*, Utah State University, Fall/Winter 2003; "In The Bottoms," excerpted from *Deep River: A Memoir of a Missouri Farm* by David Hamilton, published by the University of Missouri Press, 2001. To order this book, please call (800) 828-1894 or online at www.umsystem.edu/upress; "Rivers" by Gary Holthaus appeared in slighty different form in *Wide Skies: Finding A Home In The West*, Univ. of Arizona Press, 1998

On the cover: Whether large or small, streams and rivers help sustain life, shape the landscape, and provide an ideal place for man to come face to face with Nature. This is all true of Archer and Big Buffalo creeks in Benton County, Missouri, as photographed by Ethan Hirsh. For additional pictures, see his photo essay beginning on page 77.

To put your hands in a river is to feel the chords that bind the earth together.
—Barry Lopez

The Mississippi River, derived from the old Ojibwe word *misi-ziibi* meaning 'great river' (gichi-ziibi 'big river' at its headwaters), is the second-longest river in the United States.

—

The Missouri River, which flows into the Mississippi, is the longest river. The river is nicknamed "Big Muddy" because of the high silt content in its flow. (these two river tidbits are from *Wikipedia*)

No project is done without the help of others and this one is no exception. Thanks are owed to Shelley Oehler, Dana McGillin and Christopher Rossi at Humanities Iowa, Carole at Courtyard Books, the Iowa Arts Council, Laural Ronk with Bluff Arts, Beth Lindquist at the Union Pacific Railroad Museum, Paul Ingram and Prairie Lights Bookstore, The Presbyterian Church-on-the-River, the River Boaters of Keokuk, as well as to Thomas Dean, Patrick Irelan and, of course, each of this year's contributors. Thanks as well to Fenna Marie & Laura Lee, truly sublime water players, water creatures, water on water walkers (even if sometimes they scrurry a top of ice).

Humanities Iowa

Iowa Arts Council

Introduction:
Living In A Watery World

Native American author Linda Hogan writes, "Earth is a water planet." This is true, yet noticing that a river consists of water is really the easiest thing to understand about a river. As we focus on the Mighty Mississippi and the Big Muddy Missouri, we would do well to think of these rivers as much more than just flowing water. These two rivers sing and inspire; they flood and recede. They contain living stories and have caused all types of history. They are constructed of clouds and produce morning mist. They support a wide variety of natural wildlife and habitats both in and out of their waters. Each river resides as much in our imaginations as it does in tangible form. Being residents of this watery planet, we do our state and planet a favor by exploring and learning to understand the varied array of elements that make up these two rivers that form our state's eastern and western borders.

Contents

In The Bottoms

David Hamilton

(Note: excerpts from *Deep River: A Memoir of a Missouri Farm*, University of Missouri Press, 2001.)

Underlined Passages in My Father's *Walden*

> *The mass of men lead lives of quiet desperation. What is called resignation is confirmed desperation.*
>
> *To be a philosopher. ... is to solve some of the problems of life, not only theoretically, but practically.*
>
> *Shall we always study to obtain more of these things, and not sometimes to be content with less?*
>
> *A man is rich in proportion to the number of things which he can afford to let alone.*

So far, Father's selections are about what one would expect, and I've long assumed that the second of them was the most influential. But then comes the less expected poetry of a fifth passage:

> *It is well to have some water in your neighbourhood, to give buoyancy to and float the earth.*

When, in the mid-nineteen thirties, and working in Chicago, my father underlined that last passage, he could not have foreseen that he would become a river bottom farmer and learn a great deal about a river's way of giving buoyancy to and floating the earth, not to mention floating away many who had once lived beside it.

The Bottoms

To hit bottom. The bottom rung, rock bottom. To scrape the bottom of the barrel. If there's a top of the morning, there must be a bottom. Lake bed but river bottom, though there's a river bed too, at the bottom of the river. "Such a dream I had," says Bottom the Weaver. "It shall be called 'Bottom's Dream,' because it hath no bottom."

The river flows above its bottom, which can run shallow or deep, just as the bottom of a ship, the portion below the water line, can glide just beneath the surface or cut down deeper. Ships too are bottoms, as in "a fleet of forty bottoms." To say "that horse has a good bottom" means that it has a strong foundation and does not tire easily.

By a geological clock, the land is flowing. Sixty-five feet of it cover a limestone bed, and we call the top of that the bottoms—all the river's adjacent floodplain subdivided to plural form by the divisions of communities and circumstance of the people who live along the river. As early as 1819, an expeditionary report made under the command of Major S. H. Long remarked on the "distinctive names" already attached to settlement of the area. "We pass on the south the Chenai au Barre, Tabeau, Titesaw, and Miami bottoms." The Tetesaw Bottom, variously spelled Petitesas and Teteseau and pronounced "teetsaw," became the location of our farm.

It is a distinctive name, given also to the adjoining plain. One explanation finds it sloughed off from "Little Osage," from "Wazhazhe" or "Wacace," two ways of transcribing the name by which those people appeared to call themselves. One can imagine "Petite Wazhazhe," "Wasazs," "Sas," and "Saw" on the tongues of French voyagers and trappers. Another is "Plain des Petites Sauts," which would have derived from the numerous falls running off the bluffs that divide plain from bottom and that feed the springs. Yet another hinted at in "Teteseau," is little "kettle" or "bucket," for the ancient lakebed higher than the river bottom that became in time the prairie. So some have thought the French meant to say, "little cradle of the prairie," without trying to explain their translation, but "Little cradle (or kettle or bucket) of the Little Osage Prairie from which small falls leap down" ought to encompass the idea.

Well below that cradle, the bottoms make up a floodplain roughly level with the river bank. It is an alluvial plain formed from sediment washed in by flowing water. At times the Missouri filled the whole of the plain as glaciers melted and sent rushing back to the sea the considerable fraction of the earth's water that had been locked up in ice. Sediment the water carried sank slowly to the bottom. Then as it quieted gradually to the river we know, the Missouri wandered back and forth across the plain and kept on filling in those bottoms.

The river cut a channel across the bottoms. The river would knife along like a plow carving a furrow into the land. Plow shares are also "bottoms," with a farmer working a single-, two-, or perhaps a four-bottom plow. Prevailing winds would urge the river more toward one containing bluff line than another. Then the winds would shift and urge it back. Or once pushed and plowing deeper, another spring melt, with new rains and resulting flood, would swirl against a bank, catch on a cottonwood rooted in it, turn, and cut a new path across a bend, reforming the channel. Or sediment building up would turn rushing water away from a bench or table it had formed. Hence the rises in the bottoms, the "eminences."

Since we are a bookish people, it didn't take long for someone to imagine the river writing and rewriting its way across the land. So Mark Twain made much of learning "to read" the neighboring Mississippi. Similarly, Thoreau, early in *A Week on the Concord and Merrimack Rivers*, celebrates his farmer neighbors as writers.

> *You shall see men you never heard of before. ... Look at their fields, and imagine what they might write, if ever they should put pen to paper. Or what they have not written on the face of the earth already, clearing, and burning, and scratching, and harrowing, and ploughing, and subsoiling, in and in, and out and out, and over and over, again and again, erasing what they had already written for want of parchment.*

The figure has a long history. A chorus in *Antigone* chants its equal.

Early writing was lapidary, engraved in stone. The river, too, first carved its channel through limestone bedrock then carried in its fill of suspended, gradually settling earth. Over time, soil the river washed along settled somewhere, then often somewhere else. Soil and sometimes sand. As men and women began to map the region, the river did not always mind their maps. Parts of Carroll County to the north of the river became Saline on the south, and parts of Saline County became Carroll. A channel would shift from the right side of an island to the left, with the right side filling in until it became continuous with the bank. So we came to farm accreted land.

Much earlier, the Missouri Indians became familiar with and filled in the land, a small piece of which became our farm. Their nearby townsite covered the upland ridges, while they gardened and hunted bottomland on both sides of the river. Meanwhile their cousins, the Ioways, the Otoes, and the Osage, all Siouan language speakers, filled in to the north, west, and south. The Otoes may have entered the region with the Missouris (or Missouria as was once said), and divided from them, moving west, then north until they came to "Ni-bra-th-ka, with a trill over the 'r,' meaning flat or shallow water." So they named Nebraska for the Platte, "a mile wide and foot deep." When the Sauk and Meskwakie pushed and were pushed into the region, they contested with all four.

These nations, you could say, taking a geological view, were another sediment that built up, settled, shifted, and defended their sites of tentative advantage—as white settlers would become later. The primary sediment though, in granular form, was sand or silt or clay, with sand the most coarse and clay the finest. Sand alone isn't much to farm. Clay, though very fine in its granular form, bricks up hard as it dries and is not easy ground to work. Silty loam has more organic matter, more humus, and more sand. Our Front Field, the one bordering a county road, was mostly sandy loam. It worked easily and was less lumpy beneath your feet. If you knelt and took a clod up in your hand, you could work it apart in a moment into potting soil.

At the other end of the farm, our North Field was gumbo, a dark, clay soil that is unusually sticky when wet. Sink in to your knee and it will suck your boot right off. Common through the lower Missouri and Mississippi valleys, it gave its name to the soup, or the soup to it. The first definition in my dictionary is for the gummy pods of okra that give that soup its body, and the word is said to be of African origin, from the Bantu. My father loves okra; we had plenty in our garden. As soil, gumbo has more clay than silt and bricks up, too, leaving deep cracks in the dark earth as the summer lengthens. It was the hardest ground to work when wet, the least tractable. When the plow cracks gumbo, the earth falls away in chunks or rolls over in long, thick sheets like a quilt too thick to fold and put away.

Bottom farms don't require top farms for the sake of definition, and in fact some of the county's top farms are in the bottoms. But bottom farms do require, for the sake of definition, farms in the upland plains or on the hill. The upland Tetesaw Plain overlooks the Tetesaw Bottom. The plain, a combination of alluvium and loess, is virtually rockless and deep. The bluffs on both sides of the river, the plain itself, and the soft defining hills around it are all mantled in loess, which means that for ages we can count but not comprehend the air was often filled with earth in its finest form. For a couple of million years the soil kept building—some washed in by the river, some blown in on the wind. Farmers on the higher plain are thought by those in the bottoms to be looking down upon them.

In 1952, when levees had yet to be repaired after a major flood of the year before and so the bottoms flooded for the third time in our first five years of farming, my brother and I assured our parents that they would not have to maintain our fifty-cent weekly allowances. We stood on the bluff with upland neighbors watching an inland sea cover everything for miles in all directions except behind us. It would be another decade before I saw an ocean. We figured we could mow lawns and rake leaves for our pocket money.

"We were moving peacefully in clear, calm water," Marquette recorded in June 1673, as he and Joliet descended the Mississippi,

> *when we heard the noise of a rapids into which we were about to fall. I have never seen anything so frightening, a confusion of entire huge trees, of branches, of floating clods of earth, rushed from the mouth of the Pekitanoui [Missouri] River with such force that we could not pass without grave danger. The agitation was such that the water was all muddy and could never be clean.*

June is flood time and "a June rise" part of parlance. The explorers had only happened upon the usual. Early newspapers record flooding for almost as many years as not. In 1921, for example, "all the pecan groves in the lower bottoms were flooded and water drove the Chevalier family to seek shelter upstairs."

Later, we became the Chevaliers' immediate neighbors. Joe, the youngest son, my age, often worked a field across the road and took more chances than I with his tractor. One spring day he ventured one round too far into a swag and sank to the axle on his large back wheels. There he sat in the distance, spinning and spinning, and watching his wheels as if studying how far down he could dig. "I've heard of China," he might have been thinking. We found a log chain, hitched our tractor to his, and pulled him out. Turning to farming after the eighth grade, Joe stayed in the bottoms. During the 1993 flood, he kept his outboard "loaded flat," bringing his gunnels down to water with the last possible sandbag, and put-putted out along the levee to where they'd best be laid. He worked as if all the river were but another field to him.

In a letter to my aunt, my uncle describes the 1943 flood—only a moderate one by his records. Coming into one farm house by boat with neighbors, he found a cat "dry but weak" in the kitchen. Unc took him upstairs, set out bowls of canned milk and water and opened a loaf of bread. The cat could make it several days more "if necessary." In a second house, the men "chinned" the water as they moved clothing and a mattress upstairs. As they wrestled the mattress up and threw it on the floor, a black snake poked out his head. Downstairs,

> *just about everything was floating but the stove. A washtub was there, dry inside, I threw the pillow in that and let it float. I took off the lamp chimney and put it on a high shelf, so that if it went much farther the chimney would not stop the upward progress of the table.*

That's the uncle I remember, improvising like a sergeant-major, using what was handy to keep a pillow dry and foreseeing that the lamp chimney, catching first against the ceiling, would shatter before it could push a table down under water determined to rise. And so it went in 1947, '51, and '52 until the Corps of Engineers stabilized the river channel, raised levees, and built a series of dams upstream. Then the floods were fought and "won" for forty years.

To say this work was controversial, would be an understatement. The levees controlled many a June rise but brought flooding, when it did occur, to new heights by containing and so increasing the speed and force of the flow for as long as possible. Thousands of acres on upstream tributaries were lost to lakes behind containment dams in a system of denial to some, including Native Americans on their reservations, for the sake of barge traffic and more dependable farming elsewhere. The river served as an auxiliary railroad, and the actual railroads took quick advantage of more dependable, better controlled bottomland by laying hundreds of miles of track along a level surface. Cold War planning for better lines of internal communication was a factor. A grain surplus was an advantage. Big-time agriculture was never more positively envisioned, and big-time vacation lands began to beribbon the Dakotas.

The great flood of 1993 came on slowly. There was the usual runoff of winter melt from the northern forests and from the western mountains. Then came the rains. Spring rains turned into summer rains. All June it rained. Every other day it seemed to rain, which is probably an exaggeration. No, every day it rained somewhere on the river system. The rivers swelled and kept on swelling while rain rained on itself. Fields lay patched then smothered in ground water that could find no place to soak in. Every stream, creek, and rill became a river. Creek beds one could have stepped across rose as torrents. Satellite photos centered on Iowa showed that its density of water approached that of the Great Lakes. All the rivers flooded. But rather than cresting and subsiding, new rains sent crest after crest rolling down each stream. All along the rivers, levee systems held and held, and then were overwhelmed. With '51 in mind and so with only slight exaggeration, the '93 flood became spoken of as a 50-year flood. As it kept on cresting, it became a 100-year flood. Before it was over, we spoke of a 500-year flood as if we all had personal recollections of pre-Columbian eras, and we watched water fill the landscape as must have been normal toward the end of a glacial period when the bottoms were smoothed to ten and twelve mile widths by long term glacial melt. Driving through the bottoms later, my brother George and I found watermarks on trees two feet above the top of his pickup cab.

When my father and uncle came to the bottoms, they bought the land that was available and that they could afford. If they wasted any time wishing they farmed on the Tetesaw Plain from the edge of which we looked down on one flood after another, I failed to hear about it. Much later, with my uncle dead and my father retired, and having sold 400 acres of our 430 acre farm, we heard of an upland farm for sale and went to look it over. It was rolling, with creeks and draws working their way toward the river. It had a pond fringed with trees. I couldn't tell whether my father coveted the place, high over the floodplain, or if he just wanted to stick with farming. "We know farming," he said. His father and his grandfather had known farming. Like my brother, both had been Georges, named for those who work the earth.

At farming we were competent; that's what my uncle would have said. "Competent" was his favorite word of praise. "He's competent," he would say of another man, likely a neighbor farmer. When I finally looked it up, I found the Latin *petere*, "to compete," "to strive." A competent person strives capably against others. The root, moreover, that small part, *pet*, names feathers. Swifter than it seems, "competence" is winged and flies, or flows, along the beds of ancient rivers. So if you get right down to the bottom of it, my uncle's wry understatements could be lost on the ignorant.

Clearing

> *Notwithstanding the general prairie character of the county, a vast belt of timber land, of from one-half to six miles wide, fringes the Missouri river, and corresponding belts fringe all the lesser streams, of which there are many… In the fall of the year the vast wooded bottoms…abound in pecans, hickorynuts, and "mast" of nearly every kind. Wild grapes, summer grapes and fox grapes, flourish and bear luxuriantly.*
>
> —A History of Saline Country, Missouri, 1881

When my father and uncle were clearing, I tagged along whenever I could, which has much to do with my intermittent absorption in outdoors things and everything with my learning to use an ax. For in a frontier community such as I could pretend ours was, giving a boy an ax was a crucial step toward his becoming a man. I had a camper's ax, more than twice a hatchet but not full sized, and was learning to use it. Clearing our farm, my father and uncle used bulldozers, chain saws, dynamite, and fire, but they also used axes. Several well worn, double bitted axes were among our common tools along with a handful of heavy files to sharpen them.

Jess Miller, who worked for us, was a wizard with an ax, and I saw evidence of it. In a kind of John Henry contest, my father and uncle raced him one day, they with a two-man chain saw, he with an ax, to prove to Jess that the saw was more competent. They managed, though the awkward two-man saw could not do some things as deftly as an ax, and it took them most of the day to accomplish a lead Jess could not hope to overcome. Once or twice the chain broke and had to be repaired. They used up three such saws clearing the farm.

One of my father's claims about Jess was that he could take a tree leaning one way and drop it the other. As I remember now, the feat had something to do with a particular tree being festooned with wild grape vine and using that tangle and the tree's own weight as leverage. In *Two Little Savages*, by Ernest Thompson Seton, the story of rural youths who play Indian and learn woodcraft through a coming-of-age summer, and much my favorite book at that time, one protagonist takes up a bet that he cannot fell a tree against its lean. A two-minute time limit figures in the story. But Sam does, and Seton describes in detail his first cut and his second, then how he gets the tree to break backwards, you might say, and waits tantalizingly for just the right wind to give it an extra push so that, with his last strokes, he fells the tree and it drives a stake into the ground exactly along the line prescribed.

My own axemanship was less competent. One of our jobs each spring was to walk through new fields and cut out willow and box elder shoots emerging from land we had cleared and plowed the year before. The old woods persisted in the young shoots. Those would be limber, sprouting up crookedly from uneven soil that had been loosened when turned over and then had settled through the winter. The shoot was light enough to cut in a stroke, but only by a stroke well placed. You had to strike just beneath ground level so your ax would meet resistance. And you had to read the shoot into the soil to know the angle at which to hit. If you struck too high or missed its curve into the ground, the shoot would quake, and a small shock reverberate along the ax handle into your hand and arm, and you'd break stride and have to try again. Part of the skill meant moving through the field steadily, sizing up where and how to hit by instinct, much as a boy running along a path through the woods shuffles his stride without conscious thought to better leap over a log. The unconscious element was critical. Having a sense of what lay *sub-tela,* beneath the weaving of the soil, was subtle.

Much the same was true of splitting logs. Far better to guide the ax deftly than to whale away with it. What was crucial was intuiting where to put the bit against the log, then being able to place two strokes on the same line without thinking too much about it. It also meant keeping your ax clean and sharp and knowing how to use the file, stroking evenly toward a center point along the line of the handle. Work like that partook of romance for me, though only because I never had to do it on a daily basis. Even now I lament storm damage less than I should. It justifies getting out my adult double bitted ax that I bought for three bucks at a garage sale and restored, using oil, a whetstone, and a heavy file that I bought for the purpose and that cost four times as much as the ax. The handle was split too, and I set in a new one. Now it clips off the limbs and divides logs for the fire.

Felling trees, though, was just the glamorous part. My father and uncle bought a secondhand John Deere D with cleated, iron wheels. Slow, simple, and powerful for its time. They bought a Model A Ford and used that as their first farm car. Hitching the Model A to a wagon, they hauled to the farm an iron V that they had had welded together from two side frames of a car that had been junked. It lay on the ground like a twelve foot tuning fork and measured eight feet across the open end. They were tuning up their land, and they needed more than saws and axes. They needed a rake, the rest of which they made from cottonwood 2x10s and 4x4s sawn from oak.

They assembled it in the shade of cottonwoods on a house site abandoned before they bought their uncleared land. Setting upright a dozen 4x4s, they bolted them to an axle joining the open ends of the fork. These were to be the rake's teeth, but something had to hold the teeth to the ground. So they bolted four lighter planks at ninety degree angles to the teeth of the rake. Then they hinged a long, running 2x10 to the iron frame and held it back with a spring over the ends of those horizontal planks. The plank kept the teeth of the rake in place, dragging rather than turning while being pulled forward. The teeth then pushed brush into a windrow.

When enough had bunched up, the 2x10 could be pulled forward against the spring and the horizontal stays released so that the teeth fell backwards and the axle turned. The teeth would pivot 180 degrees until their formerly upright ends fell straight to the ground. Turning, they left the gathered brush behind. Another set of stays then reached forward on the horizontal and the 2x10 sprang back over them to steady the teeth, which could then start another windrow.

Jess had done his part first, as had a bulldozer with a shaving blade. After taking a chainsaw to the larger cottonwoods, they pulled out the stumps with a tractor and log chain. The toughest required dynamite. The rake came into play when they began to see a field. One wonders how the pioneers managed, without the oldest tractor I've ever driven, much less the bulldozer, chainsaw, log chain, dynamite, and the ingenuity of a couple of vigorous and experienced men, the younger one, my father, trained as a mechanical engineer. They both remembered smaller, lighter, mule-drawn rakes that had made windrows of cut hay.

Their implement was typical of their faith in the makeshift, in nothing's being junk that could be salvaged for a purpose, in always finding a way to invest more of themselves since that's what they most had to spend. My father learned to weld, the nearest blacksmith was miles away and a day often lost to the smallest repair. A pile of scrapped equipment mushroomed beside the machine shed, and they always searched it first to construct a needed part. It was a pile of toys no one ever made them put away. Often it proved essential.

What fully prompted their labor, I cannot know. Years earlier, on only their second anniversary, my father and mother had agreed "to slow down and enjoy life" if Father wasn't a business success in ten years. It didn't take quite that long, for eight years later, stuck in middle management, he became restless. Uncle Henry was older and had long worked around farming though not at it. Perhaps he had assumed he would take over the "home place" near Auxvasse, which his parents had lost. To begin to recover that loss, he bought eighty acres of uncleared bottom land and started again. On vacations, Father came from Chicago to help. Then the 430 acres became available and they seized the chance, investing all they had and all they could borrow.

Years before, Uncle Henry had done an unusual thing which may have reinforced his desire. Shortly after the First World War, he had worked in Europe for Quaker relief missions. Through the long winter of 1923 and 1924, on the heals of famine, he had been stationed on the Polish-Russian border, between Vilnius and Riga, and had directed efforts to rebuild villages and reestablish farms in territory that had been systematically depopulated with most villages burned to the ground. But the refugees were returning, living in hovels and burrows, and perched on the edge of starvation.

Uncle Henry worked with peasants and with veterans of the Polish and Russian armies. He started a school for orphaned boys and helped train them to farm. When his duty ended, he traveled to Moscow and decided against standing in line to see Lenin's body, the line being "too long." He worked his way south through Kuybyshev and around the eastern side of the Aral Sea all the way to Tashkent at the foot of the Himalayas. In one two week period, he traveled over 6000 miles by Russian railroad, always in third class, the level just above bare boxcar, and only took his boots off twice.

The year before had seen the worst of the famine. Writing from a town on a tributary of the Volga, he told his mother that "for two winters the wagons have picked up the dead. Men have eaten dogs and dogs have eaten men." In morning sun, he saw the pock marks on walls "made by the rifle fire of passing revolutionary and counter-revolutionary armies." He photographed the burial grounds of German, Polish, White Russian, and

Bolshevik armies and listened to witnesses from all sides tell their stories. He became cautiously friendly with the secret service agent assigned to follow him between Moscow and Tashkent and heard the agent express an interest in coming to America. Would there be farmland? He knew at close second hand the fragil and contingent nature of supplies and the vagaries of weather, all of which was reinforced a decade later by drought and dustbowl and the Depression.

He found beauty, too. Writing in his later years he described "certain sounds that will live always in our memories" and ended his Russian memoir this way:

> *It may be the moaning of the wind around the corners of the house on a cold snow-covered night as you pull the covers closer about you, firm in the knowledge that out in the barn the livestock in your keeping has been cared for.*
>
> *Perhaps it is the swirl of water as the ship cuts through the black and silent sea and the stars shine upon you. The smell of salt water is in the air, the bell on the bridge tolls the hours of the watch as you and other members of the crew sit or lie about the deck and smoke a last pipe before going to your bunks.*
>
> *But the most vivid sound of all to me is the creak of saddle leather and the steady muffled thud of horses' hooves in the snow, in a silent and very distant Russian forest.*

The first memory comes from his youth in Missouri and from the farm he longed to recover. I recognize as true to him his taking responsibility for the livestock. The second recalls his working return from Athens to Alexandria to home, a rare interlude at sea for a farmer. The third fixes that formative winter when he helped a rural people loosen the grip of famine. He was twenty-five and in charge of veterans at a work station on the Russian-Polish border. Its essential sounds echoed the knowledge of his youth and were translated over the next sixty years as the sounds of farming. In his younger brother he found a willing partner.

To Clear

To clear one's mind, or heart, or throat. To be freed from burden or obligation or from legal charge. To gain or profit by a knowable amount. To settle things, to clear them up. To own something "free and clear," neither mortgaged nor encumbered.

To come into a clearing, where the sun shines. For the extent of the clearing to be without obstruction to your view. To be free from confusion and so free of doubt. For a solution to be evident to the mind and a problem, therefore, solved.

When my father and uncle cleared land for a farm, they engaged in a foundational act. Since the eastern half of The United States was virtually all forest when Europeans arrived, almost all cities and farms, unto the Mississippi, began with axes ringing. "We are here," they sang, "making it clear that we claim a place for ourselves." It's hard to lament that overmuch when here we are—cities, commerce and industry, universities, and farms.

West of the Mississippi, as woodlands gave way to prairie and plain, much less clearing was required. Instead we think of our pioneers as sod busters, turning long grass under, and of their building sod houses. They cut what they found before them and made their first homes out of the material at hand. In Saline County, too, especially on the Tetesaw Plain, there were few trees and settlers turned farms up out of the soil without clearing much forest first. But the bottoms were heavily wooded. To farm the bottoms, timber had to be cleared. Trees, shrubs, vines and brush.

In its unflagging attempt to clear more things up, my dictionary says that the Indo-European root of "clear" leads both to "clamor" and to "claim." Clearing the land set up a roar. Axes, bulldozers, chain saws, and dynamite; what a ruckus we made. The oldest plat maps show all the county divided up and claimed by someone. But much of that was in timber, especially in the bottoms, and until an ambitious farmer came along to clear it, it may have been a deed on paper but it was much less a deed in fact.

As my father and uncle cleared their land, they began to see a farm. They could see ahead to planting and harvest because they had gained a clearer view of what they possessed, by deed, surely enough, but also, as another idiom goes, by making the fields their own. They could look across a would-be field and begin to see its shape and condition. The geography of the farm became featured by what they saw: the Strip of Timber; the New Field; the Half Circle of Timber; the Long, Narrow Field; the Dogwood Patch. I always guessed about the Dogwood Patch. I didn't help clear it, and then it was gone; and so I have always been uncertain of its place except to know that it named the southeast corner of our 100-acre Middle Field. Its features lay in the memory of my father and uncle rather than in my own, and it must have been a difficult piece of ground since they referred often to break downs on it.

My father and uncle cleared the land and made a farm from it. Later they cleared off all the debts they had incurred in its making. Thus they cleared the way for another generation. They got a lot out of their fields. They wouldn't have responded readily to "What's your field?"—the first question at a university. They could better answer what was in one—corn this year, beans last. Perhaps their usage foreshadowed some tactful doubt about possession, even though the deed was recorded and money laid down on the marble counter in the bank.

Once a New England friend complained that Midwesterners fail to hunger for "peaks of excellence." Our accomplishments tend toward the horizontal. Farmers bend to their work on both sides of the road, and river, and in the long run, health, weather, and perseverance level out much difference. I find it hard to imagine, "He's an eminence in his field," being said of a farmer. On my uncle's tongue an "eminence" was a visible rise in the lay of the land, a bench or table. It might swell up waist high across a field. Your tractor would chug a little more crawling up on it. If my uncle were to use "eminence" to describe a neighbor farmer, he would have meant only that through sustained competence, over many years, a man had pulled himself to some small advantage, which is exactly what he and my father managed. Later George joined them at it.

Which makes none of us clairvoyant, another relative of "to clear," though chiaroscuro effects they knew well. The play of light and shade all over the field as cloud shadows cross it. The interplay of willows along the drainage ditches, and through the swags, with corn in the well cleared field. Or of plan with execution, of the present with the past, of the generational and social demands made, over time, on a piece of land. Of a farm at cross purposes with wetlands, and of the Missouri Department of Conservation with us. It is probably an error to think that we ever clear things up. At best we clear the way for a spurt or two of what looks like progress, most of which calls for correction later on.

Ice

We say "river," half forgetting that we mean "bank," French *rive, ripa* in Latin. Bank dwellers, we define "river" by its enclosure rather than as the "fluence" that flows or floods past us. A fish might have done it differently, and so can winter, which arrives *("a-rive")* at riverbanks with a vengeance.

In December 1837, a year before Miami had been laid out, but by which time settlers were bunching up on both sides of the river, a Reverend Eli Guthrie of DeWitt, on the north bank, ran a ferry that joined the two places. John McMahon and Perry Harris operated it for him and both lived on the Miami side. McMahon was the son of one of the first settlers in the Miami bottom; and Harris, who was only nineteen, boarded with McMahon, his young wife and child. On this December evening, chunks of ice floated on the water, crowding at the bends, and the river was beginning to "gorge" at some unseen point downstream. The two young men were determined to cross and set out in a flat-boat large enough to carry a team of horses.

The night was cold and growing colder. In the early dark, ice chunks already seemed larger on the water. Guthrie urged the men to spend the night with him, but McMahon and Harris insisted on crossing. Guthrie followed them to the water's edge, trying to change their minds, but whether a young wife was calling, or a girlfriend, or both; or whether it was the bravado of being young and as bold as ice themselves, they refused to listen. Within yards of the shore, ice thudded against their boat. Their poles and oars became straws against the river's force, and the ice and current swept them away. Guthrie and a few neighbors ran along the bank, trying to keep abreast of disaster. They saw the boat ram against a sawyer and McMahon and Harris leap off.

A sawyer is a whole tree, or most of one, washed out of the bank and carried downstream until its rootball or cracked trunk lodges against something and sticks in the river as if rooted. It sticks up out of the current, branches extended, like a boulder in white water. This one was boulder enough for the flatboat and upended it. Poles and oars fell away, the boat rushed downstream bottom up, and McMahon and Harris jumped onto the sawyer's branches and clung there, shouting for help.

Guthrie ran back to DeWitt, untied his skiff, and with William Smith and Lilburn Barns, ventured a rescue in the dark. The time for caution had passed, or at least these men suppressed it. They made it to the sawyer where McMahon, grabbing the skiff's chain and pulling it to him like a puppy jerking on a leash, drew the bow right up on the trunk, swamping the stern and spilling Guthrie and Smith. Both men went under in a gulp and were never seen again. The boat swept away, and at the lurch, Barns, who sat forward, either leapt or was launched onto a floating piece of ice, his momentum carrying him to another, and he kept on running and jumping from cake to cake until he got within a few feet of shore where he fell in and waded out.

Marrying need to the moment, Barns improvised luckily. When he dragged himself out on the bank, he was a hundred yards down stream. His companions were lost and the men they had hoped to save still clung to the sawyer in the night.

News of the disaster spread faster than winter dark. First on the DeWitt side and then soon enough where Miami was coming into being on the south, people built bonfires of driftwood to cheer their vigil and to sustain Harris and McMahon, as best they could, with at least the thought of warmth. No one else dared launch a boat that night. But they shouted encouragement and stared as if thought alone, or the penetration of sight could tie those at the fireside to the very nearly lost. By daybreak, McMahon was dying. Harris shouted the news to shore and with it McMahon's assurance that he wished all his friends to know that he died "resigned to the will of Him who doeth all things well." Not long after, Harris called out that his friend was dead. He wrestled the body higher in the crotch of limbs and secured it to the sawyer. McMahon's relatives urged him to strip the body and to use the clothing to better protect himself, which Harris also managed.

So passed one bitter winter Tuesday night in the long history of the Missouri River. On Wednesday afternoon, men from Miami loosed the ferry on their side and tried to rescue Harris. They didn't get far before they too struck a sawyer. Their boat shipped water, and ice, and the men barely managed to scramble back. Other boats were launched from both sides with men now here, now there, thinking they saw an opening and could brave it. But one by one each faltered and turned back. A chunk of ice would strike a bow and ram it off course. Another might crash amidships and threaten to board like a blind and clumsy swimmer. Surely the knowledge of one failure after another weakened succeeding attempts. If Guthrie, Smith, and Barns had been winged by little thought, anyone who tried now was burdened by too much of it. Someone located a lump of lead, tied a line to it, and man after man tried to heave it to Harris, but he was moored beyond their best throws. Like the river itself, rescue was icing up.

Now the only hope for Harris was to endure until the river "gorged" and he could walk off over ice. Other ideas must have tempted him. He had seen Barns hop from cake to cake to safety. Clinging to his sawyer, Harris must have thought and thought about when to leap, which moment to try, which string nearing was sufficient. But initiative ebbed until what little remained froze over. Nor was he launched by accident, flung beyond the paralyzing cast of thought.

Another possibility would have been just to hop a cake and chance it. Perhaps it would run into a bank or onto others that had. Then again, perhaps he could wait long enough.

Meanwhile his family, neighbors and friends became as rooted to the shore as he was to the sawyer. They could neither assist Harris nor abandon their vigil. Death by exposure was rooted in their heritage. Sam Irvine tells of the earliest days of his family in the county, and their living "out on the prayer," exposed to the winds, and townspeople believing no one so situated could survive the blizzards. A day or so after each winter storm, a rider would venture from town to find out whether the Irvines were still alive. But the Irvines had a cabin and a fire.

For three days and four nights Harris clung to the sawyer, hoping, waiting, frozen, freezing. That last night the ice chunks began to wedge together, gorge up against whatever braced them farther downstream, and it seemed that rescuers might get to Harris. But Old Man River, or Mother Nature, blunted that last hope almost as quickly as vision of it formed. Two slabs of ice swirled together against the sawyer, "chunked up" and crushed him. McMahon's body was thrown into the river and lost like Smith's and Guthrie's. Harris's stayed wedged to the sawyer so his neighbors finally got it out.

River Horse

"A tree's just a weed if it's in the wrong place." By now you recognize the voice, and if you grant the fact of farming, my uncle had a point. Five-ten would be tall enough and no slouch in his bearing. A widow's peak and thinning hair, mostly dark until late in life, parted in the middle and swept straight back. Heavy cheeks, a mustache, leathery, well tanned skin, that which was exposed on his face and arms. If relaxed, he held a pipe. At work, he'd have a tool in his hand, a hoe, an ax, or a wrench. Usually a touch of irony twinkled in his blue eyes.

He made a play of strength, and I recall a photograph of him pitching a log onto a pile for burning. The log was six feet long and thicker than his thigh—strong man's stuff for sure, which he made seem light. But it was cottonwood, not balsa but tending in that direction, so the photograph was a joke. At the same time, Father would complain that Henry didn't know when to stop tightening a nut. More than once I had to dig around to find a foot or more of pipe to fit over a twelve-inch crescent wrench and so get leverage enough to loosen a nut that Unc had worked on first.

During clearing, a bulldozer pushed large piles of trees and brush together for burning, the cores smoldering for days. Piles would rise larger than houses, and Unc would stride up, splash gasoline around one, and toss a lighted march. Whoosh! I never tired of that moment. He'd strike the wooden match on his thumbnail and stand a few feet away while warning off the rest of us.

His care for small things was legendary, when they weren't weeds. Just as he picked up Spot during a flood and gained a dog, another time he brought home Timothy Tiger, a yellow tomcat. I've seen him swerve his pickup to avoid a butterfly. Then there was the time he found a groundhog's hole on the remnant of levee that edged our strip of timber. No longer a main levee, it could still prove useful, limiting the damage of rising water. No reason to let it fall apart. The groundhogs would have to go. Fill the hole in and they'd just dig it out again. They'd have to go permanently. We owed that not just to ourselves but to our neighbors.

Unc walked over to the machine shed and filled a gallon jug with gas. He walked back and knelt beside the hole. Carefully and lightly, with the side of his hand, he brushed away a line of marching ants and waited for them to move into the brush. Then he sloshed gas down into the hole, flicked a match against his nail, and tossed it.

"I hope the folks are home," Unc said, rising to his feet. You have to get the beat and his lingering over the Os: "i HOPE the FOLKS are HOME." We turned away and went on to other work.

Late one night, during a spring flood in the early 60s, someone found such a hole on the outer side of the levee, below the water line and not easy to get at. Several hands were needed, some to fill sandbags, others to carry them to the point of defense, still others to get into the water and place them. Father and Uncle Henry, the two oldest workers, got in the April, night time river. Father took off his clothes, figuring he'd have dry clothes to warm up with when he got out. Henry just waded in. A local radio station had come for on-the-spot coverage and a reporter stood on the levee, bending toward Henry, reaching with a microphone toward his mouth while he took a bag, went completely under water to place it, then stood back up blowing river off his mustache.

I don't know what he said since I was off in college and so not there to hear. The line that lingers in general memory is Duke Chevalier's who, standing to the side, fifteen years younger and dry, murmured, "He's a tough old sonofabitch," which was his way of saying "he's competent." I remember instead another line of Unc's, for no special occasion, so it may serve for this. "The hippopotamus is the largest, even-toed, non-ruminant, ungulate in the world, and what does it get him." River horse ("hippopotamus") that he was at the moment, he knew what he was saying as he added, "We can rent this river, but we'll never own it."

Farm

To farm is to hold onto something, and a farm is land to grasp. Its Indo-European root has to do with affirming and confirming, positive assertions beneath which one senses the infirmity of a grasp that weakens. "Farm" comes down from those old roots through the Medieval Latin *firma*, a fixed payment for rented land, a farm. In Spanish, *la firma* becomes a signature, a guarantee, a signing on.

Standing on his tractor, Uncle Henry was grasping more than the wheel. He had grown up farming. He had studied agriculture at the university. He had spent a critical year in Poland and Russia after World War I, helping reestablish farming in a broken land. He had come to Saline County as the representative of the University of Missouri Extension Service. He had worked for a bank as an assessor of farmland. But he was not then an owner of land, not an affirming farmer. His parents, meanwhile, lost their farm. Their *firma* had lost most of its value. Then my uncle and father started over, beginning where they could.

Again and again my father spoke of how much he valued "being his own boss," able to make his own decisions, and to work through his mistakes. The relief he felt for what he had gained seeped from him like new meanings from old words. But I never heard Uncle Henry express such feelings. Instead I see him standing upright, his hand firmly on the wheel, the tractor performing the primal act of breaking into the land, readying it for seed, and so acting out his affirmation over and over, as if to reaffirm his parents.

My father plowed too and described it lovingly.

> *In a good sandy loam it is just plain fun, especially when you are plowing under wheat stubble and the sunflowers have had time to grow. I never*

got over my astonishment for there they stand in front of your tractor, eight, ten, even twelve feet tall and so dense you can't see through them. But their turn comes and the plow flops them under, and you look back and not a trace of them is to be seen, aside from an occasional leaf sticking out from under a clod.

"But their turn comes." I note the satisfaction in that, and the magic, too, not the least of which was seeing that sunflower flop under the soil like any superior who had got, as Father would say, "too big for his britches."

The two of them had grown up working mules like Frank pulling single-bottom plows—that is, plows with a single cutting share. They had crossed into an age of tractors with double-bottomed plows and then larger disk-plows that moved faster, tumbling milkweed, sunflowers, and pigweed—"careless weed," we called it, and whether it was careless or could care less, I never knew—turning all that under swiftly while turning up dark, workable soil. Each spring they affirmed another season and another chance as the hard worked, forgiving earth welcomed seed once more. Plowing cleared the ground for them in more ways than one. Alongside the old prairie, it served as prayer.

The loneliness of the work adds to its meditative aspect. A medieval farmer would have rarely worked alone. The stark aloneness of farm work is recent. Through the turn of the twentieth century, Hamlin Garland wrote Midwestern stories in which roads were full of people walking to and from their fields. In a Garland story, a farmer steps from his door and within moments meets neighbors. People stride with him or pass by in a wagon. He'll hoist himself onto its back and ride along with his legs dangling over. In my youth, the likelihood of walking from the Old Fort to the farm without a neighbor coming by and offering a ride was much less than even. Old section maps indicate school sites within a mile both north and south of our farm. Schools meant families, people in the plural working the land. Most of them did not hold onto it, however. Few had much of a grasp.

Those one-room country schools have been closed for more than half a century. My father and uncle employed only one farm hand in the early days, to whom they also offered a small house. Later they only brought in temporary help, often enough us boys, until George became the third farmer. Later he and a brother-in-law worked alone—just two men and large, expensive, efficient equipment farming over 1000 acres. Any one of them standing at the wheel of a tractor was a symbol of the lay of the land before us.

I won't echo Melville and say we kept "sorry guard" although it was imperfect. In 1900 the population of Saline County was 33,000, nearly all of whom farmed. Now the population has dropped to 24,000 and half of it resides in one town. The most ecologically enlightened approaches to farming would require a fair number of people living in cities and towns to return to the fields, to reduce our reliance on chemicals and our consumption of fossil fuels. I hear no clamor to return.

Nor have I volunteered. I doubt that I would have written this book had I fully grasped the farm. Recording what I have of it, and of our neighborhood and its history, has depended on my getting far enough away to look back on it, much as Father spoke of plowing, not from the tractor seat any longer but from remembering having plowed. The jolting eases, as does the dust and tangle of weeds, the share broken on an old stump and a day lost to unmounting, welding, and remounting it. So does the anxiety of working quickly enough to plant while the weather holds and of supporting the unwieldy investment in time, seed, machinery and materials that will show no return for months if at all, much less the work in cold, gloomy sheds through the winter, coaxing old equipment through another year. "What would you do if you won the lottery?" runs the joke of farm country; "Keep on farming, till I lost it all."

My brother no longer farms, and does not miss it. Quite a few of my high school friends have left or have been forced out of farming, and I can't find one of them who laments it much. The usual response is, "Why didn't I think of this sooner?" "Why was I ever fool enough to farm?" or "Sure I'll farm again, when I retire." And though I haven't heard those words from George himself, I notice that he now finds numerous reasons, far more than are practical, to drive alone into the bottoms, by the old farm, and through the Wildlife Area, thereby taking a roundabout way to or from home.

Beyond all the economic problems, today's farmer bears the brunt of criticism for a century of questionable land use, born of another time and discipline. He stands a little like the Vietnam veteran who carries home the stigma of having participated in a questionable war. There is no question that the land has a different character from when we began. Only scraps of timber remain and most wetland is gone. Among farmers in the bottoms, as in wetlands elsewhere, farming means drainage, cutting ditches and setting culverts so the fields will dry out enough to work. Among my university friends and others, that has become a transgression of the worst order; those being the same people, often, who find transgression thrilling when it is their own adventure in science or in art.

An enormous volume of chemicals has gone into the soil. Agricultural runoff damages the ground water and there are fewer worms. Even with the minimal erosion in the level and levee-ed bottoms, much has blown onto the river, down to the gulf, not to mention what comes and goes with the floods. The river has been rip-rapped, "stabilized," channeled. It can seem a canal. Huge dams hold water back in flood season, but when a flood gets out of hand, the levees, ever higher, always stronger, lift the crests and confine water that otherwise would have spread out earlier, more widely, and with less force across wide bottoms. Levees channel and aim the river at more destructive chances.

At the same time, the Tetesaw Levee, like many others up and down the rivers, held all through the sixties, seventies, and eighties, allowing the development of more efficient farms. Fewer workers labor in the fields; many more students complete high school and college. My early school years were scarred by numerous classmates dropping out after the eighth grade, which was as soon as they could legally.

Most dropped out to farm, and quite a few other emerging farmers failed to complete high school. Neither my grandfather nor great grandfather Hamilton were high school graduates, and few adults suggested it should be otherwise. Whereas George and I assumed we would go to college, it had been doubtful for our father who dropped out of school twice. Once during high school, when heart disease crippled his father, our father, at fifteen, took over their farm. Then about five years later, halfway through college this time, he dropped out again after the tornado leveled his parents' home. Only the insistence of his mother, a Chicago woman, got him back to college to graduate in time for the Depression, and to the luck of a job in Chicago, where he met my mother and prepared, at great length and indirectly, to return to farming almost two decades later.

Mid-20th century graduation photographs from the eighth grade in country schools suggest a serious and dignified commencement. Young boys, farm hands already, stand soberly in a coat and tie, well aware of their coming duty in the fields. That could easily have been the only graduation for my father and uncle. Then through that period and beyond it, as farming modernized and those same students went on to high school and college, the cost of food as a percentage of family income, and as the percentage of the nation's work time given to obtaining it, became lower than for any culture ever, except the Lotus Eaters.

Why Geese Don't Winter in Paradise

John T. Price

"Where do the snow geese winter?" my student asked. We were standing on the edge of the lake, watching a few of the white birds glide on the black water, small, brilliant clouds against a receding storm.

"Oh," I said, startled from the sleep of other thoughts, "somewhere in the tropics, I believe. South America."

This was wrong, and another student shyly corrected me, repeating what I'd told them in class, that these geese migrate from the western coast of Hudson Bay to the Gulf Coast of Texas. That's where they spend the winter, in Texas. I quickly agreed with her and feeling the blood burn in my cheeks, stepped back from the shore.

I was embarrassed, but not for the reasons one might think. These were graduate students in my environmental literature course, and I had brought them to western Iowa's DeSoto National Wildlife Refuge, near the town where I live, to witness one of the truly remarkable migratory events in the region: the fall stopover of snow geese along the Missouri River. The few birds we'd seen were beautiful, but disappointing—I had told them to expect flocks of thousands, which is common that time of year. I had told them other things about the species as well, its coloration patterns, its mating and nesting habits, and yes, its migratory routes. I had done my research. Yet, when asked a simple question about the birds, I had without pause given an inaccurate, seemingly random response.

This factual inaccuracy is why I should have been embarrassed. Instead, it was due to being caught inside an intensely personal moment. When the student asked his question, I had been remembering the last time I'd seen geese, a few weeks prior, just minutes after my grandfather had died. I'd been thinking of my grandfather and the geese, together. I'd been floating inside that surprising connection, confronted—as I sometimes am in nature, despite the facts—with forgotten desires, with questions asked and almost answered, with promises, as yet unfulfilled, of some kind of resolution. I had, in short, been caught inside a dream.

Here's how it went: My mother calls me on a Saturday morning in October to say that Grandpa Andy, her father, will probably not live out the afternoon. She asks if I want to join them at the nursing home. At first, I hesitate—he's unconscious and even if he weren't, he's had trouble recognizing me lately. Then I feel the empty space his death will carve inside my life, my body, and I have to be near him. Steph and I rush to the car and make the trip to Fort Dodge in record time. We're not too late; he's still alive. Most of our family is gathered around the bed: Grandma Kathryn, my parents, my sisters. Behind us, slumped in her wheelchair, is Grandpa's elder sister, Esther. Her hands are folded over her waist, covering the chickadee print on her oversized pink sweatshirt. She and Grandpa have been roommates at the home for several years, but recently they've had trouble getting along. Esther has complained about his hallucinations—they're crazy, she says, though she herself has begun to tell crazy stories. In this moment, though, Esther is an afterthought; Grandpa, his dying, is at the center of our concern. His eyes are partially open, crescents of white; his breath shallow, irregular. I hear the death rattle. Someone says it's a blessing, that he wouldn't want to live this way anymore. We take turns holding his hand, the good hand, the one that keeps reaching out into space, grasping. Someone says he needs us.

We have been with him for four hours straight when my father suggests we leave to get dinner. We will be quick, nothing will change, nothing has changed. We wheel Esther up to the side of Grandpa's bed—she doesn't seem to understand what's happening—place his flailing hand in hers and leave. At the restaurant, not twenty minutes later, we get the call that he has died. Grandma Kathryn cries, says she should have stayed with him. Carrie Anne tries to comfort, telling her that when she worked at a nursing home this was how many of the residents passed away, often during the few minutes it took for a loved one to step outside to get a drink of water. It was as if they chose to die that way, on their own terms. That would be just like Grandpa, I tell her. Doesn't the caption beneath his 1928 yearbook photo read: "Muse not that I thus suddenly proceed; For what I will, I will, and there an end"? Yes, Grandma whimpers as I help her into the car, that's how he was.

I close her door, and when I turn around, I see them. The geese.

I didn't think much about them at the time, a half dozen Canadas flying low over the parking lot. Migrating geese are a common sight that time of year. And I certainly made no symbolic connection between them and my grandfather—nature had never been part of the vocabulary of our relationship. With him, there had been no camp-outs, no gardening, no fishing or hunting trips. Beyond the sunny, well-groomed slip of a golf course, he hadn't seemed to care for the outdoors. This fact, like the geese over the parking lot, had never seemed unusual. Fort Dodge, the hometown we shared, was small, industrial, a place that neither contained nor stood witness to what one might call the undeniable poetry of the wild. I'd also considered that as a young man my grandfather might have come to associate nature with the freezing, often brutal labor that the Depression had forced on him: digging ditches, shoveling coal, spreading creosote. That he might have been glad to achieve the cramped, but interior walls of a manager's office—one of the landscapes that, as a child, I most associated with him. Others included the small white living room where he taught my sisters and me how to shuffle cards, the looming cavern of the Gas and Electric service garage where he let us blast the truck horns, and the cryptic facade of the Masonic Temple where, at Christmas, he passed out toys to children in need. Grandpa's generosity and kindness and fallibility were, in my memory, mostly played out within these artificial structures of home. I had believed, as a boy, that he was indelibly anchored to them, and because I loved him desperately, to us.

I was surprised and hurt, at nine years old, when he and Grandma moved to Green Valley, Arizona. In retrospect, though, I can see the signs of the migrant spirit. His parents were Swedish immigrants, and as a boy, he liked to hop freight trains to Sioux City or Omaha to caddy at golf tournaments. Later, as a young man, he liked to drive fast, colorful cars along country roads. He liked to dance; he liked to drink, which can be another kind of absence. Grandpa was always restless, I now recall, fidgeting in his seat, soon up to play with us or wrestle. His body itself became a landscape on which we traveled, riding his shoulders, climbing his legs as if they were trees. So, it shouldn't have been a surprise that after the immigrant parents were dead, the daughter grown, the job finished, he had decided to leave Fort Dodge. Especially for Arizona, where it is always warm, where he would be free to walk the golf courses and mountain trails all year, free to dream of another kind of life. Perhaps one spent outdoors.

For a few months after they moved, that's exactly what he did. Then, in the middle of a patio bridge game—the Sonoran moon was full that evening, my grandmother told me, the quail were singing— there was a numbness in his arm, a darkness, a slumping beneath the table. In the years following the stroke, he would learn to move again, to sit, to walk, but not very far, only the couple blocks to the pool and back. Mostly, he would sit in his chair in the sunroom and watch the television or the evening light as it swooned across the Santa Ritas, even more imprisoned by place than he'd been in Fort Dodge. "Imprisoned" was the word I often used—it's how I assumed I would've felt, if I were him— but as with the snow geese, as with many of the assumptions I make about life beyond me, I was wrong. Grandpa was not imprisoned; he was just beginning to move.

On the deck at the DeSoto visitor center, the students and I passed around a pair of binoculars to look for bald eagles. During my turn, I scanned the tree line and spotted a few white heads flaring like votives against the dun branches. I passed the binoculars to the student next to me, and continued searching the skies. I caught the wobbly sliver of a turkey vulture. We are made of dust, the minister had said at my grandfather's funeral, and to it we return. *But what of the air?* I thought. More than the earth, the air caresses and enfolds our bodies; it penetrates, migrating through the lungs, becoming rivers of blood, the tributaries of flesh and spirit. The air never truly leaves us, until the end. Yet, when asked, I can barely find the words to describe it: cold and warm, dry and humid, stable and unstable. Sometimes, during the spring storms—the floods, the tornadoes—I've heard it called "crazy." But what I was reminded of in the visitor center is that whatever else it might be called, the air here is also full. Of wind, of cloud, of birds—all of it moving or migrating. Right then, the students and I were concerned with birds, and though there weren't as many snow geese as we'd hoped, there were the bald eagles and the vulture. The eagles were moving from Alaska to Mexico, while the turkey vulture might soar all the way to the tropical heart of South America. Those are the kind of grand distances I liked to imagine when thinking of migration, journeys worthy of recognition, even celebration, such as the one I'd heard about in Hinckley, Ohio, where each spring residents welcome the vultures back from their odyssey. Not quite the swallows of San Juan Capistrano, but remarkable for its singularity, here where the migrations are so prominent and the celebrations of them so rare.

Whether he noticed it or not (and I'd always assumed he did not), my grandfather was raised within this full, migratory air. Any wonder that when his body became confined to chairs, his mind took flight? It was a family friend who first noticed. During her visit to Green Valley, she told Grandma that she'd been sitting with him in the living room watching *Guiding Light* when he picked up the phone, dialed Houston information and asked if there was a listing under his name, Harold T. Anderson. A few minutes later he was on the line with a complete stranger, Harold C. Anderson, asking him if he was sure he owned the house he lived in.

"But Andy was a perfect gentleman," the friend said. "You'd never have known he was, well, *funny*."

The doctor eventually diagnosed arteriosclerosis, describing how the vessels in my grandfather's brain were hardening, restricting the flow of blood. He warned that the hallucinations and amnesia would probably get worse. When I first heard this, I felt a rush of panic. I was in grad school, and though Grandpa had lived in Arizona for nearly twenty years, though I had grown up without him, I believed he still had a significant role to play in my life. I was, for the first time, awakening to the necessary complexities of commitment—to a long term relationship with Steph, to an impoverished career as an academic, to what seemed to me an increasingly solidified, unimpressive future. More than anything, I wanted to leave that future behind and set out for the unknown country, the unknown self. I had a feeling that, of all the people in my life, Grandpa might understand this desire to fly, might even grant me the permission, the courage, to embrace it. After all these years, the stories of our lives might still entwine. But not if he forgot me.

A short time later, after Steph and I decided to indefinitely postpone marriage, I took a cross-country car trip, then booked a flight from San

Francisco to Arizona. Grandma picked me up at the airport, and during the car ride to Green Valley told me that the hallucinations had gotten worse. She pleaded with me to corroborate that he had never gone to Texas, never gone anywhere since the stroke. Together, she thought we might be able to bring him back to reality. I nodded and prepared myself to be strong, to find gentle ways to correct him. For Grandma's sake and my own.

When I arrived, Grandpa was watching television in the sunroom, zipped tight inside his navy blue sweat suit. His face had aged since I last saw it two years ago, a bit more gaunt, but it still possessed a handsome, stubborn intelligence that led me to hope things weren't as bad as I'd been told. When he saw me, he straightened his back and smiled.

"Sit down," he said, after Grandma left to cook dinner. "I have something to ask you." He reached out his good hand and touched my arm. "As you know, Grandma kicked me and you out of the house a couple years ago and into the desert. She told us not to come back until you grew up and I could take care of myself. She was sick of us, and who can blame her, the way we behave. So we decided to hitch-hike to Las Vegas, and along the way this skinny fellow, Jack Anderson, pulled over and said he was from ABC News and was looking for a good story. He'd heard we were going on an adventure and wanted to film us. I told him, 'Sure, but you better keep up—we're not slowing down for the likes of you.'"

And so my grandfather and I took off, first to Vegas, where he whipped Mohammed Ali in a street fight and earned us an invitation to join the Olympic boxing team. When President Carter boycotted the games, we decided to go to Moscow anyway and returned home laden with gold medals (we'd filled in for the American track team, as well). After the ticker tape parades subsided, we played a few legendary years of football at Johns Hopkins University. Foregoing our pro careers, we moved to Texas to enroll in graduate school at the University of Houston, where I majored in English and he majored in Acupuncture, studying with the world famous Dr. Chang.

"Now, this is where you come in," he said. "I need to get back to my house in Houston to get my needles. If I can get my needles, I can heal myself and make some damn money. Have you brought your car?"

"No. I'm sorry."

"That's too bad," he said, and fell silent. We watched television.

The next morning, I overheard Grandma scolding him in the sunroom.

"Andy," she said, "John was only ten years old when all of this supposedly happened. He couldn't have gone on some crazy trip to Russia or wherever. Besides, John knows that since your stroke you aren't able to travel. John knows that, but he doesn't say anything because he doesn't want to hurt your feelings—can't you see that? You just have to come back to reality, Andy; we all love you too much."

I sat in the bedroom, listening to Grandma slam pots in the kitchen sink while Grandpa deliberately munched his raisin bran. I felt guilty for betraying her, for not doing more to anchor Grandpa in place and time, but there was something about his story that had forged its own allegiance. In a week, I would be heading home to Iowa, to the same decisions I'd left behind. I didn't want to go back. I wanted, in one form or another, to keep moving. Forever.

"So where did we go after Houston?" I asked, quietly, when I joined him for breakfast. He grinned.

"She thinks I'm dreaming all of this up," he said, shaking his spoon at the kitchen, scattering droplets of milk. "She says I don't know the difference between dream and reality, but I know the difference. I'll tell you about a dream—just the other night I'm golfing on this beautiful course in Florida. I need to take a leak, so I walk over to the bushes and unzip. I woke up and found myself standing between the beds, pissing all over my meds on the table. Now that, goddamnit, is a *dream!*"

Inside the visitor center, the students and I studied an aerial photo of DeSoto Bend prior to 1960, when the Corps of Engineers straightened or, as the tour booklet said, "stabilized" the Missouri River channel. The photo reminded me of something my father-in-law, a physicist, once said about fractal geometry, how the most beautiful images occur inside the mathematical border between stability and instability. He showed me some of those images on his computer, brilliant blossoms of color which, though unpredictable, conjured familiar associations: trees, mountains, lightning. In the aerial photo, the unstable Missouri River looked like a fractal image, its boundaries irregular and unclear, but more interesting than the Euclidian line it had become. Back then the river, like the air, was migratory, meandering one way then another, blurring the border between states, creating sandbars and shallow wetlands and oxbow lakes. In those borderlands between stability and instability, the beautiful things did indeed occur: the blossoming of wetland flowers and grasses, the gathering of wildlife, the sustenance of innumerable birds.

Some version of that beauty remained inside the visitor center, portrayed in panoramic portraits or displayed within glass enclosures. There was a family of snow geese on a sandy bank, a lonely blue goose, a swan. There was a silver fox, a muskrat, an otter. They were stuffed and immobile, yet they still had the power to evoke in me a sense of concern and responsibility that led, as intended, straight to the donation box. But when I opened my wallet and discovered I only had a ten, I reconsidered—*too much? too little?* In the mornings, while munching my own spoonfuls of raisin bran, I'd sometimes caught a back page article on the sorry state of the Missouri, on the hundreds of thousands of habitat acres lost since channelization, on the continuing disappearance of its native wildlife. Other articles had informed me that, in the interests of preserving that wildlife, the Corps of Engineers was considering restoring some of the original flow of the river, allowing controlled floods in the spring and lower levels in the summer. If this plan were ever approved, it might be evidence that this dammed, poisoned place continues to dream. Continues to shape and absorb the dreams of others. That's another kind of power, I thought, and dropped the ten in the box.

One of my last visits to Green Valley, before my grandfather returned to Fort Dodge to die, occurred just after my wedding. Following the ceremony, Steph and I spent a few memorable days in the Sawtooth wilderness of Idaho near her home, and then caught a flight to Tucson. It was my idea. On the altar of the church, during the ceremony, we had placed a single red rose in memory of deceased relatives. These included my paternal grandfather, Roy, who had been allowed the dignity to die of a heart attack while I was still in high school, long before he might have lost his mind or our idolization. Grandpa Andy was alive, yet he was less of a presence at my wedding than the dead. I wanted him to be part of the celebration, part of its real memory.

When we arrived, Grandpa was still in his chair in the sunroom, watching television. I introduced Steph. He took her hand and softly kissed it.

"Hello, beautiful," he said. "Did I ever tell you about the time John wrestled the grizzly bear in Idaho?"

The next morning, Grandma asked if I would mind putting some of my wasted medical training to work and help Grandpa with his shower. She wanted to show Steph her roses, she said, but I knew that what she really wanted was a break. So I helped him to the bathroom and undressed him as he sat on the toilet. When I removed his shirt, I noticed how the muscles in his left arm had almost vanished; how the cavities near his shoulders—the shoulders I'd once climbed—had eroded, as deep as canyons. They seemed to measure how much we'd truly missed of each other, our bodies growing older. When Grandpa began, once again, to talk about our Idaho adventures, I became impatient and a little angry. I hadn't come all the way to Arizona for this. This time, I had come for something real.

"Grandpa, I was only in seventh grade," I interrupted. "There's no way I could've put a full nelson on a grizzly, let alone snapped its neck. I was a puny kid, anyway—*remember?*"

"Well, you grew up in a hurry that year."

"I didn't grow up that fast."

In the shower chair, he began the story once again. I tried to interrupt, but he took hold of my wrist and squeezed it surprisingly hard.

"Now, hang on," he said. "Just *listen,* for once!"

He explained that, in fact, he had lived two lives. In the first, he'd lived in Fort Dodge, Iowa, and worked for the gas company and moved to Arizona and had a stroke. The adventures he was talking about occurred during his second life, his "second chance" as he called it. As he told me this, I washed his rigid, useless left arm, his wilted torso, and concluded, finally, that the opportunity for any kind of meaningful connection with this man had vanished a long time ago. Beyond his physical presence, he wasn't capable of giving me much. I, on the other hand, could at least give him my silent attention.

"You think you were puny, John," he continued, "but what I keep trying to tell you is you weren't. You were huge, seven foot umpteen inches. Of course, I was eight foot myself, but believe me, you were something to behold."

He'd told me this before, but now I was listening in a different way, my hands on his naked body, an intimacy we had never known. Minutes earlier, I'd felt his nearly dead weight against my shoulder as I assisted him into the shower. He was a large, heavy man, but I knew from photos that he'd been a small boy. I'd been small, too, an experience that had cost me a lot of adolescent anguish. Had my grandfather noticed? Is that what his dreams were about? Not just the athletic disappointments of being short (Olympic boxing?), but also my failed desire to become a medical doctor (Johns Hopkins?), my drift toward writing and literature (English at Houston?), my fear of committing to Steph (the bear in Idaho?) and, of course, my desire to leave home. It wasn't just me; I recalled others in my family who had visited Grandpa in the desert and returned with crazy, but personal stories: Carrie Ann, who had fought her way through school and a disastrous relationship, was a general in the Army; Susan, the animal lover, owned a lucrative farm; Allyson, the youngest, had climbed to the top of Kilimanjaro. Mom had become thin and healthy and rich thanks to Grandpa's acupuncture sessions. The truth of these stories, though factually inaccurate, had finally reached me. All along—was it the letters, the phone calls, the terrain of our faces?—he had understood the insides of our desires, had cared, had wanted to help. He had been far away, and yet he had known me. Still knew me.

"You wrote it all down," he said. "It's going to be your first book. You're going to make millions on the movie."

"I hope so," I said, rinsing the soap from his body. "And who'll play our parts?"

"Ah, hell," he said. "We'll play ourselves."

I wonder if, while growing up in Iowa, my grandfather ever came to know the Missouri River, its earlier, restless self. He would have crossed over it in those rickety box cars when he was a boy. I wonder if, many years later, he looked at the river from the window of the jet that finally brought him home, from Arizona to Fort Dodge, after he broke his hip and Grandma could no longer care for him. If he did, he would've seen a much different river, the one I was now standing beside, straight and fast and cold. My students and I were walking along the levee, reading how the Corps built it in 1960 to create DeSoto's oxbow lake, the lake the geese come to every year. Though the brochure didn't say so, it was obvious that the new river was no longer an option for the geese—if they landed there, they would be swept away like squawking kids on a water slide. The lake is all they have. "Maybe," a student thought aloud, "if the Corps does change the flow, there'll be more places for the geese and other birds to rest along the river." We nodded and I tried to imagine that beautiful, immense return, but inevitably, I was drawn back to the river as it is. Unable to move, it digs down into itself, its banks standing sliced and exposed like butcher's meat. The Missouri has been declared one of the most endangered rivers in the nation, which is another way of saying it is in the act of forgetting itself. This amnesia—that we allow it to happen—is one of the reasons why I'm sometimes ashamed to say I live here, why I'm still tempted to leave. By moving, the fantasy goes, I might avoid witnessing the destruction of the places I love, the places that created me. I might even find a place where people live a different, more enlightened story on the land. At the very least, I might avoid moments like the one on the levee when, against our silence, the river seemed to amplify the collective judgment: You do not deserve what little wildness you enjoy.

But what other story is possible? No matter where we live, our touch seems to automatically incur shame and forgetfulness, whether it be the natural course of a river or the aging lives of the people we love. That's what those care facilities are all about, no matter how clean or well run. As soon as Grandpa was wheeled into that room at the Marion Home, he began to truly forget. Within a few years, he had lost all of his first life and most of his second. They were replaced by more immediate, but less inspiring stories about residents hitting him in the head with chairs. For a long time, he hung on to the house in Houston, but eventually even that dissolved, and I did not miss it. His stories became, for me, less about freedom and more about the pitiful restrictions of his life. But I'd forgotten what I'd supposedly learned while bathing Grandpa in Arizona: to listen carefully to even the most restricted spirit. Even when, during one of my infrequent visits to his room in Fort Dodge, he said, "It's good to be home," I did not hear what he was trying to tell me. Instead, I focused on the limited view out his window, the screams of the insane, the pungent odor of disinfectant and urine, the imminence of death. Despite all those years wishing my grandfather had stayed in Iowa, I had, in truth, always imagined him dying in Arizona, his eyes on the mountains, the still lucid desert, his spirit on the move. This was not where his journey—or mine—was supposed to end, in the very place where it had begun. A migration, after all.

"How is it you thought the geese wintered in a tropical paradise?" another student asked as we walked back to the visitor center. She was smiling, teasing—the autumn wind was cold. I shrugged:

"You know, ancient Europeans thought geese wintered on the moon—at least I'm closer than they were."

But what I really wanted to say was: "That's how far I'd go, if I were a bird. If I were a man's dream."

When we returned to the care facility, the room seemed cavernous without the sounds of Grandpa's dying. We gathered around his bed and laid our hands on his body, still warm. Mom picked up his left hand, the bad one, no longer twisted and cramped. For the first time since the stroke, she was able to place her fingers inside his and draw them to her face. "Look at that," she said. "This is my father."

We were each given a few minutes alone with his body. During my turn, I circled the bed, looking for the rise of his chest, the twitch of his mouth and arms, the thump of his heart through the pajamas, realizing how active even his dying body had been. Inside that stillness, I had the distinct sense, as people often do, that the person I loved was somewhere else. I knelt beside him and prayed that wherever that place was, it was far away from Iowa: Houston, the Sonora—Paradise. Far away, just as I thought he'd always hoped to be.

"Did you hear the geese?"

I turned and saw Esther—she'd been there the entire time, unnoticed in her bed at the far end of the room, covers pulled up to her chin.

"Pardon?" I said.

"Did you hear the geese?"

I told her I'd seen some geese above the restaurant parking lot.

"I heard them," she said, her voice almost too low to understand. "Harold always used to call me, in the fall, when he saw the first geese flying south. He'd say, 'Sister, did you see the geese?' Just yesterday, he said that to me. He saw them out his window."

Muse not that I thus suddenly proceed. Yet, that's exactly what I do back at the DeSoto parking lot. My students have already left without seeing the big flocks. I am ready to go myself, but have paused to admire the fiery scallops of the disappearing sun. This is the time of day when my grandfather died, holding hands with his sister. He chose, if there be such a choice, to end with her, in that place, in this season. He had noticed the geese every year; had loved in his own way the cycles of the land. It held, still holds, the restless spirit that defined him. Even the sky—I can already see a few stars. Soon, if the evening is clear, I'll be able to spot Orion and my favorite, Vega, a blue-white star much bigger than our own. Vega used to be the North Star, and it will be so again in another twelve thousand years. It will guide, once more, the transient and the lost, including this sun, this planet, which are moving toward it. We won't arrive for hundreds of millions of years, but still, when I look at Vega through my binoculars, I think I see, among all the stars, a special brilliance. The fire that lights the end of our story.

So, it will be dark soon, but not yet. The brilliance of the day seems, in fact, to be growing, covering the sky with gold. Near the zenith, there is a smudge of shadow. It becomes larger, noisier, more irregular. And then they are above me, the snow geese, thousands of them returning from the fields, blotting out the sky just as I have always imagined and hoped they would. I am sorry my students aren't here to see it—will they believe me when I tell them?—but I am here, and I am happy about that. Happy, despite everything, to be home. Like he was. I wonder if that would surprise him, to know that in my memory he is forever linked with these birds, that they have become for me his life's articulation. That wherever else we might travel together, he is also here with me, watching the geese return to the Missouri River. I stare into their loud, shimmering depths and lose my bearings—who is moving, who is still? Who is alive, dead? Can anything be truly lost or forgotten? I look for the answers. They fly on, blurring, as they always have, the border between fact and desire. Rivers upon rivers, flowing forever.

Now that—I can almost hear him say—is a dream.

Traces

Lisa Knopp

Shortly after 11:00 on a prickly, hot Saturday morning in July, my brother John, sister-in-law Kim, my children, Ian and Meredith, our beagle puppy, Belle, and I arrived at Keokuk National Cemetery. I'd been to this cemetery as a child and once as an adult but had never noticed the field planted with neat rows of white identical headstones or the granite obelisk topped with a figure of a Confederate soldier. Because of its position near the junction of the Des Moines and Mississippi Rivers and as Iowa's southeasternmost town, Keokuk was a transportation hub during the Civil War. Iowa's first troops were mustered from there in 1861, and Mississippi riverboats brought the sick and wounded from southern battlefields to the seven military hospitals established there by the federal government. The Union, Confederate, and unknown soldiers who died at one of the hospitals or en route to Keokuk were buried in this cemetery, one of the twelve original national cemeteries. Now the remains of over 4,600 American soldiers and many non-military people with local roots are interred there. July 21, 2000, my maternal grandmother was joining them. We were there to put my grandmother's remains in the ground. Absent were my mother, father, brother Jamie and his family, each of whom had their own reasons for not being there.

When we arrived at the gravesite, we found a T-shaped strip of bright green carpet covering the ground. Where the two sections met sat a low bench with a polished wood top. The legs of the bench were covered by dark green velvet curtains hung on rods. The man from the funeral home sat a black box, as tall as and a little wider than a jumbo box of cornflakes, on the bench. The gold label on the box read, "Cremated Remains of: Viola Mae Parris." Behind the bench and the black box was a pink granite headstone with PARRIS carved on the face. On the left were etched the scantest details about my grandfather's life: "Arthur E./ June 15, 1906/ February 2, 1990." On the right, my granny's information: "Mae V./ August 15, 1915." Beneath that was a shiny blank rectangle where "July 11, 2000" would be carved. To the right of my grandparents' graves was Granny's sister's headstone: "Pearl E. Parker/ 1906-1989. I had to remember that her birthday was February 11, since it wasn't carved in stone and I didn't know who else would remember. And it was through the remembering of details and what they evoked that I could sustain some shadow, some trace of those I loved who had left this life and of myself at an earlier time. The man from the funeral home set the display of red roses and pink carnations that my mother had sent next to my grandparents' headstone. The gravedigger, a large, amiable man carrying a shovel, asked if we were ready to inter the remains.

"Not yet," I said. "We're going to have a ceremony."

The last few times I telephoned my grandmother at the nursing home, she told me that she was dying and that she was scared. Though she had numerous health problems, she wasn't dying of a disease. Rather, she was just old and worn out and terrified. The last time I telephoned, she couldn't do anything but moan and call my name: she had already left me. When she lay dying during the summer of 2000, almost 900 miles separated us, and my income then was too small to justify buying a plane ticket for something I wasn't sure about—or so I told myself at the time. Now I regret that I didn't go to her during her final days and brush her hair, hold her hand, and gather her stories of her childhood and her family's origins.

Of all of my grandmother's wide-ranging evocations of her past, the ones that interested me the most, the ones that left me wanting more, were set in Montrose. The stretch of river between Montrose and Keokuk is one of Iowa's most spectacular landscapes. There the Mississippi flows between limestone bluffs three-quarters of a mile apart with little bottom land on either side. And there the Mississippi flows over a riverbed comprised of almost horizontal limestone ledges: the 11 treacherous miles of the Des Moines Rapids. In the nineteenth century, travelers heading up the river stopped at Keokuk, disembarked, moved their goods from large to smaller boats and barges, reboarded, passed over the rapids, and then stopped at Montrose or Nauvoo, Illinois, directly across the river from Montrose, to transfer the goods and passengers back to the larger vessels. This laborious process spawned various businesses to serve river travelers and crews. Consequently, Montrose prospered and grew. But in 1877, the Army Corps of Engineers excavated a deep canal around the rapids on the Iowa side of the river. In 1913, a private company built the hydroelectric dam at Keokuk, which raised the water over the rapids and flooded the riverfront and the fields where Granny's father's people had grown melons in the sandy soil near Montrose. Each of these "improvements" at once made the Mississippi more navigable and caused a decline in the population and the economy of Montrose. During my childhood, Montrose was but a good place to fish or to buy fish bait or smoked carp wrapped in butcher's paper. But in my grandmother's Montrose stories, river waters receded, and the town filled again with river boat captains and crews, fishermen, bootleggers, fiddle players, midwives, jilted lovers, button-makers and tenant farmers.

One of the more tantalizing stories Granny told was about her Aunt Mary (Whitaker) and Uncle Ab Bunker. While walking down the street in Montrose, Ab had been shot and killed, leaving his wife and their children to fend for themselves. Not long after Ab's murder, one of his children died of starvation. In response, Mary, whom Granny portrayed as tough and enterprising, began making and selling whiskey, the bottles of which she hid beneath the floorboards in her house. Thereafter, her children were never hungry. Was this story, or any part of it, true? If Granny had been a child when she'd known Mary and had heard these family stories, what had the ensuing decades done to her memory of them? Since my thirties, I've been aware of my own memories fading, tarnishing, decaying, or listing a bit, the joints no longer square. First names only. Bare-bones plots instead of fleshed-out stories. The troubling sense that I've confused or misplaced details that I once knew. Or that in order to fill in gaps, I have imported new knowledge into old memories. How much frailer, cluttered and tainted was my granny's memory after all those years. Putting aside questions of reliability, what in our conversation had provoked her to call forth and impart to me this story about murdered, starving, bootlegging kinfolk, and, more importantly, what meaning did she attach to it or want me to extract from it? Even if the memory were entirely fabricated, the answers to these questions could be revealing.

Long ago, I attempted to verify the details in this story about the Montrose Bunkers. Shortly after Granny told me about them, she and I walked the Montrose Cemetery in search of Ab's and Mary's graves but couldn't find them. The cursory search I undertook at the State Historical Society of Iowa for the story of Granny's uncle's murder revealed nothing. I concluded that Granny had told me a good story with little basis in fact. But almost six years after Granny's death, when I set out to research the place that her family came from for this essay, confirmation came. On June 18, 2006, 98 years to the day after Granny's favorite aunt became a widow, I discovered a web site for Montrose that lists those citizens whose name appeared in *The Montrose Journal* between 1889 and 1926. And there they were: the Bunkers, Mary and Ad (I had misremembered or misheard his name). Through the handful of brief articles, I learned that on June 18, 1908, 60-year-old Zack Andrews shot and killed 40-year-old Ad Bunker. Andrews was declared insane (he told the court that the Lord had ordered him to remove all Democrats from the earth) and committed to the state asylum in Mount Pleasant. What a tumultuous time that must have been for Mary. She and Ad had seven children. Ruth Viola, named, I suppose, after Mary's sister Viola, was only a month old. Eleven months later, Ruth died. A little over a year before the murder, Ad and Mary's daughter Ann married Zack's son, James Andrews, whose tuberculosis was the source of a complicated conflict between Ad and Zack. James died a little over a year after his father-in-law's murder, leaving Ann with two small children. While these details confirm the factual basis for my grandmother's story of Ad and Mary and suggest a factual basis for her other stories, they do not tell me what I really want to know: why Granny loved Mary so; how Mary survived the loss of her husband, daughter and son-in-law; how the river looked and moved and smelled in those final years before the dam was built. At the time of Granny's death, I had none of this corroborating evidence from *The Montrose Journal*. I hoped that at her funeral, I'd meet people who could confirm or deny the existence of Mary and "Ab" and others who appeared in Granny's stories and offer me their own memories of the people and places of which she had spoken.

But there would be no funeral. According to my mother, Granny had paid for a burial but not a service. I shouldn't have been surprised. For as long as I can remember, Granny had something against funerals, though I never learned what. She and her brother Ginger had made a pact that they wouldn't go to each other's funerals. Consequently, she stayed home on the day of his service. When my Great Aunt Pearl ("Pertsie") died in 1989, my grandfather in 1990, and my Great Uncle Buster a few years later, Granny boycotted those services, too. Some say they don't go to funerals because they want to remember the dead as they once were. But this is a flimsy excuse. For most of us, our most intense period of encoding, the process of turning what we perceive into memories, occurs between the ages of 10 and 30. Psychologists call this an "autobiographical memory bump." Consequently, most of us have more and stronger memories of parents, grandparents and siblings from this period than any other. While I remember my loved ones' aged bodies and circumscribed lives, while I remember them in nursing homes, hospitals, or coffins, in my sharpest, most indelible memories, my parents are always in their thirties or forties, my grandparents are always in their fifties or so, and my brothers are always children or young men. Surely a glimpse of one of her siblings embalmed and coffined wouldn't have weakened my grandmother's memories of them as young, robust, and engaged in the world.

Years before she died, Granny told me that she wanted to be cremated. I told her that I did, too. Cremation seemed so civilized. Flesh, fire, ashes and pulverized bone rather than a corpse pumped full of formaldehyde. Yet I felt uneasy about this request, as if it were something I should have protested. "Please don't. I can't bear the thought of your body being turned to ashes." But I didn't. Nor did I ask the right questions: "Do you want your ashes scattered? If so, where? And if your remains are scattered, what of the empty place in the cemetery between your husband and your sister?" Apparently, Granny had told my mother or someone else of her request, because her cremated remains were sent from the funeral home in Ohio near my parents' house and the nursing home where she'd spent her last few years to the funeral home in Keokuk.

What makes me uneasy is this: I hold the far-fetched and macabre hope that if memory is an input that causes enduring changes in one's neural tissue, as neuro-psychologists claim, then perhaps in the future we can exhume a loved one's embalmed body and harvest the engrams or traces, the fingerprints of stimuli that serve as the basis of memory. But fire makes dust and smoke of the autobiographical memories borne in the hippocampus, the cortical structures surrounding it, and the neural pathways that connect them to the cortex. Gone forever are the memories that one might retrieve from those interior places.

I placed an obituary in *The Gate City Daily* in Keokuk, where my grandmother was born and lived most of her life, and *The Hawk Eye* in Burlington, Iowa, my hometown 45 miles up the river from Keokuk, where she lived from the time of my grandpap's retirement until she moved to Ohio. The obituary reduced my grandmother's life to two tidy paragraphs. When I sat down to write it, I had all of the facts that I needed but one. When and where had my grandparents married? August 3, 1932 in Kahoka, Missouri, my mother answered. I knew the rest of the story. They had crossed the state line, the Des Moines River, into Clark County, Missouri, where they eloped. Granny had been 16, Paps 26. After the ceremony, Granny returned to live in her mother's house, as if nothing had happened. A couple of weeks later, her mother got wind of her daughter's marriage and sent her to live with her new husband. When Granny was 17, she delivered twin boys, Jackie and Jerry, both stillborn. When she was 20, she gave birth to my mother, Patricia Mae.

In the obituary, I requested that condolences be sent to me at my home in Lincoln, Nebraska, in hopes that I'd hear from people whom my grandmother had known. I intended to write back to each of them. But Granny had lived away from Keokuk too long to have friends or co-workers there who remembered her or felt connected enough to send condolences. She was the sixth of seven children, all of whom were dead. My grandfather had died ten years earlier. He was the second of thirteen. His few surviving siblings were very old, and I didn't know where to begin looking for them. My mother was Granny's only child; my two brothers and I her only grandchildren. Only one person, the wife of one of Granny's nephews, wrote to me, and that was several months after Granny's death. If we had had a funeral, would anyone have come?

July 15, John called to say that he wasn't comfortable with the idea of strangers putting Granny's ashes in the ground with none of us there. "Would you meet me in Keokuk?" he asked. Apparently he needed a ceremony, too.

We stood in a semi-circle around the box of Granny's remains. "Let's each tell a story about her," I said.

John nodded.

When we drove past Montrose that morning on the way from Burlington, where we were staying, to Keokuk, I remembered that Granny and Paps had taken Jamie and me to fish for bluegill in the lilypad-covered Mississippi River sloughs. "I'll start," I said. "Granny was the first woman I'd ever seen who wore blue jeans. She wore them with one of Pap's plaid shirts when she went fishing, which was often. That was before everyone started wearing jeans. Granny didn't mind being different. In fact, one time she told me about a patient at the hospital who told her that she was weird. Granny said to the woman, 'I like you, too.' With that attitude, where might she have gone in life if she'd been educated?"

I asked Ian what he remembered. At fifteen, Ian was the eldest of Granny's six great-grandchildren and the one with most memories of her before she became blind, hard of hearing and immobile. "When she babysat me, she made 'quick pizzas.' A piece of bread with catsup, oregano, cheese, and bologna toasted in the oven."

"She made me those when I was little, too," I said.

"Her house was always dark," Kim said.

"I think that's because she used to work nights at the hospital nursery. She kept the drapes closed and slept on the couch during the day. Maybe she grew to like it that way."

"I remember her running the Seventh Street Hill in that little Rambler of hers," John said. "She'd speed up the hill—we'd hope to heck for a green light once she got there or no traffic if the light turned red—and then, because there was a bump at the top of the hill, we'd be airborne."

"Yes!" I said. "Our stomachs would drop. We thought we were flying. As we climbed the hill, we'd shout at her, 'Go, Granny, go!' like the Beach Boys'song."

"The Beach Boys?" Meredith asked.

"'The Little Old Lady from Pasadena.' 'She drives real fast and she drives real hard/ She's a terror out on Colorado Boulevard .... Go Granny, Go Granny, Go Granny, Go,'" I sang.

John continued. "One day—I think I was alone with Granny—we went flying over the top of the hill and there was a cement truck. Parked. Filled the windshield. Let's just say that Granny had very good reflexes."

"I remember when you guys took her truck away from her," Kim said.

"She was driving too fast," I said. "Running up on curbs. Taking corners too sharp."

"Scaring children," John added.

"I remember when she started watching *Oprah*," I said. "After she'd seen several shows, she reached the conclusion that 'people used to have problems, but now they just have stress.' Profound, huh?"

"She never said my name right," Ian said. "She always called me 'Eeeen.' So did Paps."

"Yup. Always drawn out like that, too," John said.

"One time she wanted to call Ian to do something," Meredith said, "and she got mixed up and said, 'Jamie! John! Shit! Eeeen!'"

When they heard our burst of laughter, the man from the funeral home and the gravedigger who had been chatting beneath a shade tree stopped and looked at us.

"When I was born, she wanted my name to be Melissa so she could call me Missy for short," I said. "I used to be embarrassed by that name and told her to call me Lisa, but whenever she tried to say my real name, it never sounded right. Kind of strained. Unnatural.

"Eventually, I decided that I liked being called Missy and wouldn't have minded at all being named Melissa. Any mail that Granny sent me was addressed to Missy Knopp. When I was pregnant with you," I said to Ian, "Granny and Paps suggested that I name you either Robin or Sally."

Ian grimaced.

"So what's the deal with her names?" John pointed to the box of ashes, which said "Viola Mae," and the gravestone that said "Mae V." At that moment, I realized that I was the repository of stories and facts about my grandmother's life. I was the authority.

"She was born Viola May, but flipped it. Granny changed the spelling of May to Mae, so there wouldn't be any letters hanging down below the line when she wrote her name on her supplies at the hospital. I guess she thought that an 'e' was neater than a dangling 'y.' I have the scissors that she used to cut dressings off of patients. The blades are bent at an angle, and one blade has a knob on the end so the patient wouldn't get jabbed. There's a tag in a clear plastic ring on the scissors with her name on it. No loops hanging down." Suddenly, I wondered about this story. My mother's middle name was "Mae," with no "y" hanging down. Granny didn't start working at the hospital until my mother entered high school. Had Granny been named Viola Mae at birth, or had she exchanged "May" for "Mae" years before she ever went to work in a hospital, say, around the time of my mother's birth?

The gravedigger and the man from the funeral home were watching us, as if they wanted us to finish. I had one more story to tell.

"When I was in my last semester of college at Iowa Wesleyan, I had a tuition bill that I couldn't pay. I'd already taken out a $2,000 guaranteed student interest loan and was paying off a car loan. Somehow I had to come up with another $1,000 if I wanted to graduate that May. But I had no money. If I had put off graduation until August, it would've been too late for me to get a teaching job for the fall. Graduating in August wasn't an option, either, since I'd still have had to come up with the money by mid-July. I told Granny about my problem. That's when she and Paps were living in that tiny aluminum trailer at Mom and Dad's place. Paps loved that trailer because, theoretically, he could put the wheels on it anytime he wanted and hit the road. But Granny hated it—the cramped quarters, the shoddy construction. 'We're just baching it here,' she'd say. Or was it Great Aunt Pertsie who said that?"

John shook his head. "Somebody said it."

"A few days later, she gave me $1,000 out of her meager savings account, and I graduated with my class. She never asked for the money back. But twelve years later, when she didn't have enough money to buy groceries, I paid her back—$40 a month for a couple of years."

The gravedigger approached. "I hate to hurry you folks, but I've got another burial this morning. Will you be much longer?" How odd this ceremony—five people laughing, talking and laughing—must have seemed to him.

"Just give us a few more minutes," John said.

We prayed; then I snapped pictures of my grandparents' gravestone, Granny's remains, and nearby family graves. John signaled that we were ready for the burial. The gravedigger moved the box of ashes and the bench and rolled up the green carpet that had covered a small rectangular hole. He sat the box of ashes at the bottom of the hole. We may have sprinkled it with handfuls of earth, though I can't say for sure. Then the gravedigger filled the hole and patted a square of sod in place. It was done. Our ceremony of remembrance had been intimate, spontaneous, funny, and respectful. And for the duration, it had raised the dead. I believe that Granny was pleased.

The graveside ceremony was over, but something felt unfinished. Rather than driving back to Burlington, we bought food at the Hy-Vee deli and drove to Rand Park for a picnic. I do not remember going to this park as a child, though I do remember Granny, who owned no property other than her and Paps' slim plot at the cemetery, driving near the park along Grand Avenue, where Keokuk's wealthy early citizens had built big elegant houses near the edge of the bluff above the river. We'd marvel at what I now recognize as Tudor, Queen Anne Revival, and Colonial Revival mansions. Granny often pointed out a mansion that I remember as being on the river side of the street (actually it was several blocks away, I recently discovered) reputedly built by slaves and later used to hide runaway slaves. A plausible story. Unlike Iowa's other southernmost counties, whose lower border was a straight human-drawn line, the southern edge of Lee County was a wiggly line drawn by the meandering Des Moines River. Consequently, Lee County dangled like the loop of a "y" into Missouri, once a slave state. And it was a statue of a Confederate rather than a Union soldier that kept watch in the cemetery.

At the entrance to the park on a high wooded bluff above the Mississippi rose a statue of Chief Keokuk, a member of the Sac and Fox (Mesquaki) tribes who had farmed and hunted on both sides of the Mississippi. Keokuk wasn't the rightful leader of the two tribes. When Black Hawk, who had a better claim to the title, sided with the British in the War of 1812, Keokuk refused to join him. To some, Keokuk was a traitor; to others, he was a realist. On behalf of the US government, General Winfield Scott appointed Keokuk chief of the two tribes. Thereafter, Keokuk ceded his people's lands to those who were squatting on or attempting to buy them, and, under his leadership, the Sac and Fox, along with several other tribes, traded their lands in the upper Midwest for reservations in what is now Kansas. Keokuk died there in 1848. In 1883, the people of the City of Keokuk had the body of their namesake exhumed from his Kansas grave and reinterred in Rand Park, reputedly on the site where the "chief" had held his council fires on the edge of the bluff. In 1913, the Daughters of the American Revolution marked his grave with this statue, now green with age. Stern-looking, heavy-browed Keokuk, wrapped in a blanket and topped with a war bonnet, gazes across the wooded bluffs across the Mississippi and into Illinois. In the winter he appears to be watching the hundreds of bald eagles that ride the thermals, that fish in the open waters near Lock and Dam 19, or that perch or roost in the trees on the Illinois shore.

In my living room, I've hung a framed black-and-white photograph of my grandparents and mother standing by this monument. My mother wears a big-brimmed hat, frilly dress, tights, dress shoes and no jacket. She says that her mother liked to dress her in yellow to complement her red hair, so I imagine that her hat and clothing are a very light yellow.

She looks to be about four, which means that my granny, who is wearing a floral print dress and a long plain jacket, hose and pumps, is about 24. She wears wire-rimmed glasses; she's parted her hair on the right, lightly curled it and pulled it back; she's pinned a tiny dark hat to the left side of her head and a flower to her lapel. My grandfather, who is about 33, wears dress pants, a white shirt, an open sweater with a zipper, and a cap. Judging by my mother's outfit, her lack of a jacket, and the bare branches, I guess it to be Easter, and they've just come from church. My mother stands on a ledge on the Chief Keokuk monument between her parents. Granny has shifted her weight to one leg; Paps stands with his feet comfortably parted. But "Daughter," as Granny always called my mother, stands primly, with both feet together.

When Ian was three, he, my friend Pat and I stopped by Rand Park, studied the statue of Keokuk, and gazed upon the swift shining river. But on the day of my grandmother's funeral, I felt that I was seeing the park for the first time. We explored the flower beds, the playground, and guessed where in this park Mark Twain, a printer for *The Keokuk Post* during the 1850s, had delivered the 1886 Fourth of July address.

"You guys want to run the Seventh Street Hill?" John asked as we were cleaning up after lunch. Of course we did. We drove down Grand Avenue and followed Seventh Street, also US Highways 61 and 136, past the courthouse and across Main Street. As we were climbing the hill toward the intersection of South Seventh and Timea, we saw the traffic light click to green. John floored it; I did the same. What if someone traveling the opposite direction made a right turn on red as we were speeding up the hill? What if a cop saw us? Did Granny ever think of such things? I watched John and Kim's car disappear over the hill. "This is it!" I shouted, as we crested the hill. Ian and Meredith were expectant. Nothing. Nothing but a smooth ride down hill. At the bottom of the hill, the intersection of Seventh and Palean, John yelled out his window, "It's gone. They've fixed the hill. We can't fly without Granny." With Granny's joy at being airborne, perhaps we should have scattered her ashes from a bluff above the river.

I motioned for John to follow me and turned onto Palean Street. For 12 or 13 years, my grandparents lived in the house at 512, the longest they'd lived together at one address. What may have accounted for this longevity is that they'd worked out a good deal with the landlord: instead of paying the rent, they paid the yearly taxes on the property—or some such arrangement. But, too, they may have stayed because they liked the place. As a child, I spent plenty of time at 512. When I was four, my father was transferred from the CB&Q locomotive shops in West Burlington to the freight car shops in Havelock, Nebraska. A little over a year later, my young and homesick parents decided to return to Burlington. Since it was several months before we could move back into the house that we had rented out during our absence, my father lived in Burlington with his mother; my mother, Jamie, and I lived in Keokuk with Granny and Paps. (John's birth was several years away.) Once we'd moved back to Burlington, Jamie and I spent a couple of weeks every summer, occasional weekends during the school year, and part of our Christmas vacations at 512. As I drove down Palean Street, I was awash with memories. Was it 1965 or 2000 or both at once?

On the day of Granny's funeral, Palean Street was quiet. But when I was a child, it had been otherwise. My grandparents had lived next door to Elders', a store owned by Wendy and Helen, with an apartment above where Wendy, Helen, and their son, who swept floors and helped at the meat counter, lived. Elders' reputedly had the best meat in town, sold Dixie Cream doughnuts, Quincy Maid potato chips and cottage cheese in tall drinking glasses; smelled of ripe bananas, worn wooden floors, and raw meat; gave credit to their neighbors. Granny would run into Elders', grab a carton of eggs or a can of soup and flash it at Helen, an enormous woman in a dark print dress who sat on a stool behind the front counter. Helen recorded what Granny had taken; at the end of the month, they'd settle up. Granny said that when workers went on strike at Hubinger's, Elders' gave them credit for food but not for beer and sold the catfish, game, or homegrown vegetables they brought in. But what I remember most vividly were the rows of glass jars filled with penny candies on the shelves behind the counter. Helen waited while Jamie and I stood before the counter, taking forever to decide. Then, per our request, she'd drop Pixy Stixs, cigar-shaped bubblegum, wax lips, atomic fire balls, jawbreakers, Tootsie Rolls and little malted milk balls into brown paper bags so tiny that she could only get the tips of her plump fingers into them. We paid for the candy with Granny and Paps' empty pop bottles.

Years later, when Granny heard on the radio that Helen had died, she told me stories I'd never heard before. Wendy had fed meat scraps to Paps' rabbit-hunting dog, Belle. If Wendy ran out of fresh tomatoes in the store, he'd reach over the fence and pick those that Paps was growing. Though this exchange seemed to favor my grandparents, Granny seemed irked by Wendy's theft of their tomatoes. And she remembered that Wendy had bought his building for a song during the Depression from a religious group that was awaiting the end of the world. When the prophesied day of the Rapture came and went, the group sought to regain their sanctuary. Wendy refused their appeal.

For me, Elders' was the heart of the neighborhood, though others might have placed it farther south. At the intersection of Palean and Progress-Commercial Streets, a tavern occupied each of the four corners; a half block to the west sat another tavern. Paps was a regular at a tavern owned by his friend Howard, who lived in a duplex a few doors down from 512. Often, Paps helped out behind the bar. Pop Whitaker, my great-grandfather, was also a regular there. On the day that his social security check arrived, he'd ask my Great Aunt Pertsie to drive him to Howard's, where he'd cash his check and buy drinks for himself and the "blisters" at the bar, as Pertsie called his once-a-month women friends. When Pop's money was gone, he'd call Pertsie to take him home, where he'd wait until his next check arrived.

Howard's was the only tavern in the neighborhood that I remember entering. Recently, I asked my mother about it. "When you went in," she said, "the bar was on the right."

"No, it was on the left, I said. "The pool tables were on the right."

"No, it was on the right. I can see it now."

Maybe she and I entered by different doors. But as I mull over the differences in our memories, I do see the bar on the right and the jukebox at the end of the bar nearer the front door, where I'm now convinced they belong. Why had I moved the bar to the opposite wall? Had I confused Howard's with some other place? I try to remember the names of the other four taverns. The Shamrock? The Buffalo? I remember with certainty the Black and Tan, the only one of the taverns whose parking lot I could see from the front bedroom window when I was supposed to be sleeping. Watching the men enter and exit the building and mill around in the parking lot both frightened and excited me.

My grandparents' house was in the middle of everything; the neighborhood I lived in in Burlington was downright dull in comparison. People walked past 512 to and from Elders' or the taverns or the Cozy Corner where we ate hamburgers and drank milkshakes. It was near enough to Hubinger's, the corn starch and corn syrup plant, that there was traffic to and from the factory. By mid-morning each weekday, the odor of the cooking corn enveloped us. I breathed shallowly until I became dull to the heavy, foul, almost yellow air. (I smell it now as I write.) The tow-headed Stimpson boys, who lived across the street from 512, stood on their porch roof and held contests to see whose arc of urine reached farthest into the front yard. The grass had been mowed, but the flower bed was thick with foxtails and crabgrass. Granny had grown flowers there, if not every year, at least once. I reach hard but can't retrieve the name or color or shape of the flowers. Zinnias? Petunias? And why when I looked at this weedy patch did I have a wisp of a memory of a bright rooster, Paps' friend Duke Dexter, and Granny shooing Jamie and me into the house because of something Duke was doing or saying that she didn't want us to see or hear?

Perhaps this and other memories are lost forever because the relevant traces or engrams have blurred and weakened. Or perhaps traces of everything we've ever experienced are still there, adrift in deep, dark waters, but we can't retrieve them because we don't have the right cues, the right bait. Or perhaps we forget not because memories are lost through decay or displacement but because the human head and heart only have room for so many stories and images of the past. Because the formation of new memories interferes with the retention of old ones, the brain erases the old memories as it lays down new ones. By erasing memories of my childhood, I have more room for memories of my children's childhoods. By erasing memories of my grandparents, I have more room for memories of my parents. But I suspect that the process of deleting and retaining memories is less logical and exact than that.

As a child, I thought that Keokuk and Palean Street ended at the same place: about a block south of 512. Then, I had no sense of beyond. But my map of Keokuk shows Victory Park separating Palean Street from the Mississippi, which curves around the bend where it meets the skinny little Des Moines River. What this map leaves out is that a bluff separates the old neighborhood and the riverfront park, just as forty years, more or less, separate me from the time when most of my memories of this place were laid down. Despite the passage of time, despite the changes in the house and street, this is where my grandparents live in my memories. Paps is an ironworker who spends his days off fishing or hunting and too many nights as a customer or bartender at Howard's. Granny is a nurse's aide who works nights in the nursery at the Graham Hospital, pampers her mean Siamese cat, Jezebel, pulls catfish from the river, and has remarkably good luck running the Seventh Street Hill.

It isn't enough to leave my children with photographs and written anecdotes about my grandmother's life. Rather, I want them to commit to memory at least some of these fragments: that their great-grandmother, with her minimal education, read voraciously; was always the first to leave a family gathering, and while she was there, preferred to eat alone in the living room rather than at the table with everyone else; loved to dance; never scraped a butter wrapper ("We're not poor!" she'd say if she saw my mother or me do so); smoked cigarettes but quit; never cooked a meal that anyone raved over (her cooking, like mine, merely suffices); disapproved of pierced ears (only women like Paps' sister Eslene, whose daughter ran off with a race car driver and left her kids with her parents, had pierced ears); never allowed a political discussion in her presence for fear that people might "get mad or feel abused"; was blind as a young child because of something that Pertsie had put in her eyes; told me what to expect when we got to heaven (angels, harps, lovely clouds); spent her days off from the hospital at the river fishing but gradually spent more and more of her days off over-medicated with a sick headache; cut pizza into slices with a pair of scissors; was violently allergic to the mayflies that that rose in clouds from the river each June and invaded our riverfronts for a few days; taught me how to draw a design that her mother, who ran a chuckwagon on the Oklahoma range for a few years, learned from an Indian (or so the story goes); kept secrets, though at times leaked tantalizing details; would rather hold a baby than do just about anything else; delighted in those who defied the rules (a woman who bootlegged whiskey in order to feed her children; a wealthy, local citizen who hid fugitive slaves); never turned me away or denied me anything.

In truth, it's not so much my grandmother whom I seek to impart to my children but me. After all, my memories aren't replicas of events but subjective, emotion-saturated records of how I experienced those events. I want my children to carry some of these fragments within them to guard against that day when my memories have been whittled down to nubs, the traces eroded to nothing or gone up in flames. It's a selfish motive, I admit. Yet, too, if my children know something about the place-times that I cherish and those who inhabited them, they'll understand why I am as I am, which in turn will tell them something about themselves. If they know this, then there's nothing to withhold or forgive.

After taking pictures of 51 Palean (the "2" was missing), we found the hospital where Granny had worked and the school where I'd had the same kindergarten teacher my mother had had 21 years earlier. Often, Granny met me after school in her white nurse's uniform with a piece of cake from the hospital cafeteria. On our way out of town, we passed by Lincoln Elementary. Because I've never seen a picture of Granny as a child, because I don't remember her telling me stories about her time as a student (did she ever get in trouble or win a spelling bee?, did she eat lunch alone or with a gang of kids?), I cannot see her there. We paused in front of Great Aunt Pertsie's house on South Seventh Street, within walking distance of the confluence of the two rivers. There Granny lived as a child, long before there was indoor plumbing, long before someone sold off most of the land. Our ceremony of remembrance was complete.

After we returned to Burlington from Keokuk, Ian and Meredith swam in the motel pool. I sat nearby on a chaise lounge and jotted down details about the stories we told and the sites we visited. But as I write about this weekend almost six years later, I realize that I did not record enough about my children's responses to the day. And, now, that matters as much as my attempts to evoke, thicken, and vivify my memories of my grandmother or of myself at an earlier time.

I spread out the photographs I took of that weekend. On the day of Granny's funeral, Ian wore knee-length denim shorts, a black hat (a fishing hat to my eyes), and a short-sleeved shirt. In the photograph, the shirt looks solidly gray, but I know that the pattern is comprised of tiny squares of black, white and gray. In a few weeks, Ian would turn sixteen and enter high school as a sophomore. It would be his last good year before things fell apart. Nine-year-old Meredith wore a sleeveless white dress, drawn at the waist, and with white embroidered flowers covering the bodice. Her hair was pulled back in a bun; her skin was bronzed from daily swim team practice; she wore her new wire-rimmed glasses. When I asked her what she remembered of that day, she spoke in detail of her dress, which she loved; riding to the funeral in John and Kim's car; and stopping at McDonald's on the way to Keokuk. She vaguely remembers being at the grave but doesn't remember Palean Street, Rand Park, the river or running the Seventh Street Hill.

If Meredith rode with John and Kim, then Ian, Belle, and I rode to the funeral together. What had we talked about? Ian remembers singing the

Queen song "Another One Bites the Dust" and me telling him that that was an inappropriate song for such a day as this. His memory stirs the soup, shuffles the deck, and, yes, I do remember him sitting in the front passenger seat singing. And I remember that Montrose, the wide river, and the bluffs on the Illinois side were in the background. Ian remembers that Belle, the dog he'd waited for all his life, the dog who bore the same name as each of Paps' and Jamie's rabbit-hunting beagles, went to the funeral and stayed in the motel with us. And he tells me that, at the funeral, he'd told a story he'd written for Granny years earlier. When she lost her front tooth (a fact), she protested her condition by rallying the other nursing home residents around her (a fiction). After they conquered the nursing home with their canes, they set out for the mall, where the residents defeated the mannequins, again with their trusty canes. Through it all, Granny was their fierce leader. As Ian spoke, I remembered the story and how Granny had laughed when Ian read it to her. But between the time that I left the cemetery and sat down by the motel pool to record my memories of the day, I had forgotten it.

Now I have a shadow of a memory of Ian at Granny's gravesite telling his fabulous story. Perhaps since he and I talked about what he'd remembered and I'd forgotten, and because we wondered out loud about the whereabouts of our copy of his story, the incident is deeply etched in his memory. Someday when my memories are frailer and dimmer and I've forgotten the presence or the whereabouts of this essay about my grandmother, an 11-mile stretch of the Mississippi River, and the nature of aging memory, I'll ask Ian about Granny's ceremony of remembrance, and he will call forth the story of his story. Rather than feeling reassured and content, I'm uneasy. What else am I on the brink of forgetting? What else sleeps beneath the surface of consciousness, waiting for the right cue to bid, lure or startle it into awareness?

Sources:

"Andrews Hearing." *The Montrose Journal.* June 25, 1908:1.

"Andrews is Insane: Had His Hearing at Keokuk Before Commission Thursday of Last Week." *The Montrose Journal.* July 9, 1908: 1.

The History of Keokuk Web Site. http://keokuk-ia.com/history/index.html

"The Keokuk Tourism Web Site." http://keokuk-ia.com/history/index.html

"J.A. Bunker Shot: Zack Andrews Shoots Him Through the Heart." *The Montrose Journal*. June 18: 1908: 1.

Keokuk National Cemetery. V.A. National Cemetery Administrative Web Site. www.cem.va.gov/nchp/keokuk.htm

"Montrose: A Look into the Past." Rootsweb (sponsor). http://freepages.genealogy.rootsweb.com/montrose

Praeger, W. E. "The Birds of the Des Moines Rapids." *The Auk* XLII (October-December 1925): 563-577.

"Taming the Des Moines River Rapids." U.S. Army Corps of Engineers, Rock Island District. 2004 www.mrr.usace.army.mil/Brochures/TamingtheDesMoinesRiverRapids.asp

Thanks to Tanya Boltz at the Keokuk Public Library, who found the articles for me from *The Montrose Journal*.

The people and places in this essay are real, however, some of the names and places have been changed to preserve the privacy of friends and family.

THE GREAT RIVER

Homage to Par Lorentz

Robert Wolf

NOTE: The Mississippi River seems to me much more of a spirit, a symbol, a power than simply a body of water. Because of its many layers and levels of meaning, because I feel it to be a symbol of America itself, the prose poem rather than the essay was the more natural form for my contribution. The present work is a draft of a much longer prose poem in progress.

Hear the song of the river ... listen ... to our story.

This is the song of the Big Muddy, the great river running from the North Woods twenty-three hundred miles and more to the Gulf, drawing waters east and west from half a continent.

Rivulets and creeks, streams and rivers ... from the Alleghenies and the Rockies they tickle and run to the great valley ... The Illinois the Rock the Ohio the Arkansas the Missouri ... beginning in the mountains of Pennsylvania and pulsing across Ohio ... starting on Colorado peaks and winding through Kansas Oklahoma and Arkansas ... pouring from Montana south across the Great Plains through Nebraska to Missouri.

The music of the names of the rivers—the Platte, Niobrara, Elkhorn and Yellowstone ... the Red, Cimarron, Canadian and Ouchita ... the Allegheny, Clarion and Youghiogheny.

Imagine yourself a thousand miles above the continental United States.

Observe the deserts and mountain ranges, the miles of checkered corn, wheat and bean fields, the forests and rivers. The Mississippi stands by itself like the stem of a giant plant, branches reaching out and up across half the continent, draining lands from the Alleghenies to the Rockies.

ii.

What is the Mississippi? A roiling mass of water and sand, earth and water … not one stream but many twisting coils moving in layers … each layer and coil unpredictable. Not one current but many … not one drive and speed but many.

The twisting coils swirl … swirl into giant eddies. Mr. Humphreys said he saw one eddy "running upstream at seven miles an hour and extending half across the river, whirling and foaming like a whirlpool."

Ropes of water snake southward, turning back upon themselves one hundred eighty degrees with a force increasing at turns. As the great snake bends, its coils and currents drive down, twisting out holes a hundred feet deep.

But the river is a series of shallows, too, deep pools and shallows down river and up, its serpentine coils increasing their drive as they move from shallows to pools and back.

Eleven hundred miles of river rushing hurtling southward … Beginning with North Woods rivulets trickling into streams gathering together more gathering and more, half the continent draining into the Mississippi, more than streams now rivers the Ohio the Missouri the Arkansas always flowing ceaselessly flowing down to the Gulf.

Eleven hundred miles of roiling water . . .into outlets, onto bottom lands … but ever the drive down the main channel.

iii.

Stand on the bank as the sun rises over the bluffs at your back … listen … feel the presence … the sound of the constant flow … the wind ruffling grasses, leaves … watch … the dawn mist rising … mist twisting into form, into sinewy threads into fibers, inchoate form into god form.

In the days of the plentiful the river yielded a vast harvest of fish and clams … in the days when the god was whole … As the god was in the days when the river was free, before the locks and dams, when the Mississipp' flooded the bottom lands of the North, leaving rich soil, flushing the earth ... In those days the river spawned different breeds of fish, and bigger ... The men who built the locks and dams told of fish as long as their boats, twenty foot sturgeon ...

And before then the Red Men of Iowa told of a time when the river was so thick with fish they could walk across on the bodies of mudcats. The days of the wild river were the days of the raftsmen and keelboat men ... wild men ... Like Mason and Harpe on the Ohio, who set up killing lots on the riverbank ... would put up a sign reading "Tavern" and raftsmen wanting a shot of liquid fire would stop ... and disappear ... throats slashed, pockets stripped and rafts and keelboats looted, bodies thrown in ravines.

iv.

In those days was a wild breed of men ... in the days of the plentiful ... in the days when the river ran wild ... when a man in buckskin would kick the dirt off his boots, stare the village loafers in the eye and announce: "I'm from Pike County, Missouri, gentlemen. They call me the Rip Roarer."

Who is the backwoods hero? He is the best at his trade ... competent, plain, practical ... half horse and half alligator.

When not hunting Indians, bear or turkey he raises hell itself, can drink three jugs of liquor and still whip five men single-handed and sing a prayer over their fallen bodies. Brave as a whole tribe of screaming savages he can winter alone in the woods, grin the bark off a tree, and wade the Mississipp' without getting wet.

He fought with Andy Jackson at the Battle of New Orleans and put Old Hickory in the White House ...

When the woods got civilized he skinned out for the rivers ... the Ohio and Mississippi ... taking cargo south to St Louis and Vicksburg ... down to New Orleans ... But by 1822 his day was over ... The river filled with the sound of the churning paddles of steamboats ... the clank and chunk of their engines ... The day of the raftsmen and keelboats was over ... and the riverman moved West ... into the mountains ... a trapper and trader.

Do you remember Mike Fink? He was the greatest of the rivermen, a hunter and Indian fighter with a sure fire aim and an arm as steady as a rock … lithe and sinewy as the great river itself, another Hercules, king of the keelboat men … king of the river.

Imagine a barroom of rough cut lumber filled with unwashed bodies smelling like swamp rot and dead hogs, bloated, flesh ready to pop, with flies swarming. Add the stench of month old liquor fumes and clouds of pipe smoke and you've got the picture and scent of a river bar. Now imagine Mike Fink, ten yards tall, stepping into this human hog house, standing in the doorway, scanning the human refuse, sizing them up, then announcing:

"Gentlemen! I'm tellin you I'm a Salt River roarer! a ring-tailed squealer! a reg'lar screamer from the ol' Massassip! I can lick five times my own weight in wild-cats, and I can use up Injens by the cord. I can swallow children whole—raw or cooked. I can out-run, out-dance, out-jump, out-dive, out-drink, out-holler, and out-lick, no holts barred, any white thing in the shape o' human that's ever put foot within two thousand miles o' the big Massassip. Whoop! holler, you varmints!—holler fur the Snapping Turkle! or I'll jump right straight down yer throats, quicker than a streak o' greased lightening. . . I ain't had a fight fur two days an' I'm spilen' fur exercise. Cock-a-doodle-do!"

At seventeen an Indian fighter living alone in the woods, living on dried corn and jerky, a crack shot from the Alleghenies … a feared man, an Indian killer. They said he could shoot a tin can off someone's head at a hundred yards.

When the settlers began crowding up the forests, Mike made for the river … and with the advent of the steamboat, took work with the Mountain Fur Company as a trapper and hunter. Worked the country round the mouth of the Yellowstone, near Fort Henry.

So it has been from the beginning … Even the gods must die. The old ones pass, and as they pass beneath the world's horizon, new ones arise. Fink's time had passed. A fame that blazes like tungsten, falls, loses luster, and sputters … Mike was drinking heavily … his old brag now pathetic …

v.

It was the old story: man versus machine. And every time a man or woman went up against one, they lost.

Except once ... Remember John Henry? Muscles like steel cables, black as ebony, the best in his line. Went against a machine. Heart mind and muscle against mindless metal and steam. For once mindless metal lost but the black giant burst a vessel and died. The one man who did beat a machine ...

Switch scenes. Camera on the great river. Trees on the bank. We hear the chunk of an engine behind a bend. ... see a cloud of smoke above the trees ... a boat comes into view ... puffing smoke ... big paddlewheel alongside, churning the water.

And that was the end of the keelboat.

The steamboat came to the river in 1814 and little by little in years that followed pushed the keelboats and rafts off the river.

The first one, the "Clermont," built in Pittsburgh by Robert Fulton had a low pressure steam engine, a side paddlewheel and the hull of an ocean liner ... She came down the Ohio to Cairo, Illinois ... and from Cairo to New Orleans. Made it south despite the great earthquake that sent the Mississippi flowing backwards. With the wrong type of hull she was punctured on a stump and sank.

The next steamboat on the river, the "Washington," had a rear paddlewheel and flat bottom with shallow draft, perfect for the sandbars, tree stumps and snags on the river. In 1815 she made the trip from Louisville to New Orleans in five days, a trip that took the keelboats nearly forty-five.

The muscle strain and sweat of the keelboat man polling his boat upstream against the river's twisting coils ... straining twisting muscles of forearm and thigh, wrist and back of the men fighting the twisting layered coils ... the complicated interlock of pistons and rods, boiling water pumped into a steam vat driving the pistons ...
the turn of the engine ... and the crash and the crash and the crash of the pistons the rods the coils going dah-dah-dunk dah-dah-dunk dah-dah-dunk ...

The steam pressure turning pistons driving wheels churning water
First one big paddlewheel then two big paddlewheels
pushing wheels churning brown water into yellow foam
sun dazzle on droplets & sprays thrown
by the rolling paddlewheel ...

Now passengers can cruise, sleep and dance on the river while Mike Fink lives in a cave off the Yellowstone.

The dangers of the river, sandbars, snags and shifting channel make the steamboat pilot a hero … navigating down river and up, noting the sandbars and shallows, aware of the shift of the river, shift of the channel …

The pilot's the hero, a man with a long memory and careful eye … What is Mike Fink now but a reflection of memory.

vi.

Mike Fink, drinking heavily, quarreling … holed up for the winter in a cave off the Yellowstone with a store of whiskey and a youth named Carpenter. What was their relation? Rumors spread and distrust between them arose.

Fink had changed. No longer ten yards tall … Sharp eyes dulled … the massive arms limp with hanging flesh … shoulders humped … This the man who in former years could freeze a room to silence. Still … there was the occasional flash …

And now the snows passed. The trappers at Fort Henry held a frolic with plenty of whiskey, wrestling, river yarns, songs, and dances.

Fink proposed he and Carpenter erase suspicions with a shooting match—each to shoot a can off the other's head. Carpenter shot first, and wham! the can flew off Fink's head but the bullet creased his scalp.

"I didn't expect so poor a shot," said Fink, and leveled his rifle as Carpenter placed the can on his own head. Fink drew a bead and fired, and Carpenter fell dead.

No one said anything then, but everyone muttered it was murder. Fink retired to his cave but even there heard the accusations. Those of Talbott the blacksmith were loudest and Fink threatened to kill him. Talbott was ready.

And when Fink, rifle on shoulder, came to the smithy, Talbott seized two pistols and called: "Stay there."

"I mean no harm," Fink said. "I came to convince you I didn't mean to kill young Carpenter."

Pistols raised, hands shaking, Talbott repeated, "Come no further," but Mike Fink lifted a foot above the threshold and Talbott fired both guns, two balls landing in Fink's broad chest, staggering him backwards. Eyes wide and mouth agape he fell.

The river god is not a god but a withered drunkard shot by a coward.

vii.

The dead god returns to the river … nearly forgotten … but here …

They are all here, all who once lived or worked or died on the river … crowding the banks …

Nothing dies. Neither thoughts nor acts … So it is with the murder of nearly nine hundred fifty Sac Indians, starving squaws children and grown men some murdered first in Illinois, most in Wisconsin on the banks of the river.

Start with this:

A band of Indians led by Chief Black Hawk on the Wisconsin shore of the Great River holding a stick with a white flag, facing a U.S. Army steamboat. Blue coats with rifles line its sides. Moments pass. The air is dead still across a space of twenty yards, until a white calls out:

"Sacs or Winnebagos?"

'Sacs."

The dead air jumps alive with a shot. The Indian and flag fall.

Like a carnival shooting gallery live indians move back and forth as blue coat sharpshooters pick them off. And were it not for the crack of rifles the Indians might have heard the sizzling that preceded the canister's explosion. Twenty-three hit the ground, dead or dying.

This is August first 1832. Flip back the calendar to April that year, when a band of one thousand Sacs leave their Iowa hunting grounds and cross the Mississippi, fleeing famine … running to Illinois to plant corn.

I came to the massacre site one chill May evening … to hear fish breaking water, to watch the currents coil and rope, to enter the silence of Time past. But a chain saw rips the air … idles … then rips again. A towboat upriver sounds its deep horn. Rock music blasts from a radio. Close by a dozen mobile homes and a hot dog and souvenir stand line the bank. I sit nearby, waiting for the spirits of the dead to rise like steam from the river.

It was the old story: whites wanted the land, all of it, for themselves. When the Sac arrived in Illinois, settlers murdered some and the

Sac retaliated. And the Illinois militia, including Abraham Lincoln, chased them into Wisconsin.

Black Hawk set an ambush ... killed a few ... and the band split, some going southeast for safety to Fort Crawford on the Mississippi ... but were rubbed out by soldiers sent from Fort Crawford. Black Hawk and his band pushed on west a day ahead of the militia and at the river began frantically building canoes and rafts. But before they finished, the Army steamboat had paddled up river and faced them as they stood on the bank. While the militia pushed through the swamps and pinned them from the rear.

Twenty-three fell. Those who escaped that first round of slaughter run for the river and swim towards islands twenty yards off. Most are shot and killed, thrashing in the water. Those who make the islands are hunted down by soldiers. Those who manage to swim the Mississippi are met their enemy the Sioux. Of the thousand Sac who left Iowa in April, less than one hundred fifty survived the Black Hawk War.

A fisherman drifts towards shore, radio playing quietly. He does not know that he is drifting over the spot where one hundred twenty-four years ago hundreds of dead Sacs floated down river in a mix of blood and water.

viii.

The river is a record of succession ... The White Man rubbed out the Red Man ... the steamboat finished the keelboat ... the railroad killed the steamboat ... now the railroad and towboat run neck and neck.

The people are forever inventing ways to make more for themselves ... Wanting more farmland in the Valley we destroyed wetlands and forests ...

Wanting to tame the upper river with locks and dams we lost the giant sturgeon and mudcat. Man is a hard master, the people are always wanting.

In the days when the river ran wild, before the locks and dams, the bed of the upper river was thick with clams ... Clam shells for buttons cut in factories on the upper Mississippi ... for buttons sent across country. And so we harvested clams ... until the clams were gone.

Before the locks and dams the north country was covered with forests. Forests for lumber. Lumber for homes and barns on the prairies! Lumber for Chicago, Milwaukee, and St. Louis! for the cities burgeoning with immigrants! for the new towns of the West growing out of mud and sod!

Lumber camps sprouted in the forests of Wisconsin and Minnesota, big crews handling saws and teams of oxen, cutting the old growth from the North Woods, spawning tales of Paul Bunyon and Babe the Blue Ox ...

Big men in the cities owned the lumber camps ... owned the forests, hundreds of square miles of old growth doled out to them by the government for putting through the railroads ... and they clear cut the land for seventy years, sent logs down tributaries to the Mississipppi where they were lashed and nailed together into giant rafts and sent down river to towns like Dubuque and Rock Island, to Cairo and Quincy.

Rafts seventy yards long with three or four thirty-foot oars at the end to guide them down river ... and longer rafts the length of an ocean liner and twice as wide pushed by steamboats.

Lumber for the new towns and the farms! Lumber for the burgeoning cities!

The people are always wanting. Land and water, lumber and wheat, fish and clams. Wanting, always wanting.

ix.

Mr. Eads built himself a diving bell to locate steamboat wrecks. James Eads learned the river like no man before or since, knew it from the bottom for fifteen hundred miles from Iowa to the Gulf.

In the darkness ... at the bottom ... blind ... Eads felt himself sucked into the bed as the current pushed, slammed, lashed ... struck like a whip with sudden violent shocks ... always the constant unending pulse of the river.

The sand at the bottom, he wrote, drifts like a dense sandstorm ... "At sixty-five feet below the surface I found the bed of the river ..." His feet sunk into the bed ... Current as swift there as at the surface.

No better place to sense the power of the river than at Keokuk. Walk out the long cement causeway to the great building housing three turbines ... three huge rotating spinners forty feet across, moved by the strength of the current, creating electric power for miles around ... The sound of the turbines is deep ... vast ... The unseen force of the river manifest in the sound of the huge rotating turbines.

x.

The currents at the bottom of the river are as swift as those at the surface ... A roiling mass of water and sand, earth and water ... not one stream but many twisting coils moving in layers ...
Running over the clear cut lands of northern Minnesota ... the clear cut lands of Wisconsin ... a sheer path for the runoff from the slopes of the Iron Range ... coming down Wisconsin fields and hills ... down every creek and rill, every stream and river running to the Mississippi.

For years experts predicted a devastating flood ... and argued how best to prepare ... with outlets or levees ... either allow the flood to escape into channels and wetlands ... or raise dikes for a thousand miles, from Cairo to New Orleans.
The levee people said, "Contain the river. It will scour itself, deepen its own channel." And with men and mules, mules and men, levees were raised for a thousand miles ... from Cairo to New Orleans.

In 1927 the waters came. Down the Red, down the Cimarron, the Platte, the Rock, the Ouchita ... down Allegheny, the Ohio, the Missouri, the Arkansas ... The river rose and more men were needed to raise the levees ... And even as the waters smashed the levees, blacks continued piling sandbags ... under guns and rifles.
The river tore at the levees ... found the weak spots ... lashed at them ... scoured them out ... tore them up ... The levees could not hold! Flood waters covered the tops of towns, farms were washed out, crops and livestock drowned.

xi.

So we put in locks and dams ... from Minneapolis to St. Louis ... twenty-nine locks and dams forming twenty-nine pools ... and a nine foot channel maintained by the Army Corps of Engineers for the safe transport of wheat and oats, soybeans and barley for the merchants controlling America's grains.

The keelboats are long gone, and with them the men who could pole
forty-five days upriver from New Orleans to St. Louis. Instead of
keelboats and rafts the upper river in summer is filled with water-
skiers and motorboats, and men and women fishing for carp and cat.
Still, there is hard work ... worthy labor.

xii.

All day thunderheads have piled up in the west.
All day fish have crowded the bank of the slough.

Now it is night ... sheer black.
Above the quiet slapping of the river comes the soft putt of a motor and
a running wake.
Three men in a launch slip along the bank,
Three men not speaking.

The launch angles as the slough twists.
The launch tows a flatboat carrying a seine.
They ease to shore ... tie the launch ... move to the flatboat and unfold
the net ... stake the seine like a tennis net across the slough, bottom
line weighted with lead, top line with cork.
One man stays with the net standing in mud, waist deep in water.
The others move upstream. One in the flatboat strikes a plunger into
the water. Fish thrash in panic, boiling the water. The standing man
waits.

Lightning rips the horizon. Thunderheads explode. The kneeling man
jerks. The plunger strikes the water ... again and again. The fish race
faster.

Oarlocks creak as the flatboat nears the standing man.

The night sky is ripped again by a zig zag of fire. The corkline goes taut.

The standing man clambers into the boat. The three struggle
to pull the lead line across the slough as it rises and plunges with
trapped fish. The water roils faster, splashes everywhere.

One retrieves the launch, slides it alongside the net and pulls the corkline over the boat as another begins dipping the fish net. And together they pull the seine tighter yet tighter, forcing the fish into a mad thrashing ball.

Every sinew every wire of muscle in their arms backs and legs ache, yet they continue pulling and dipping and hauling, pulling dipping hauling until four thousand pounds of carp and buffalo are pulled and dipped into the launch.

To the north clouds slide and part and the nimbus ghost light of the quarter moon spreads a luminous smear over the launch and the men as they angle down the twisting slough to the channel and the lights of town.

xiii.

When towns and cities and farms needed lumber we clear cut the North Woods. When the people wanted polished bits of shells for buttons, we cleaned out the clam beds. When the farmers coming to the Valley wanted land they drained the wetlands and clear cut the forests of the Delta and sent millions of tons of soil into the river.

And in the rolling lands along the upper stretches of the river straight line plowing up hill and down sent millions more tons of topsoil into the upper Mississippi. Now soil compaction caused by chemical farming causes runoff that erodes the land, sending millions of tons of topsoil into the river.

Now the great river branching like a plant across half a continent is filled with toxic fish and farmland poisons flowing to the Gulf … poisons that have created a Dead Zone.

xiv.

A fat man with a coast-to-coast megaphone, venomous spittle flying from his lips, denounces the stewards of lands and rivers.

The fat man is one of the people, born and raised in Missouri, on the banks of the Mississippi. He knows the folk, plays them like a violin, his own people …

And when he strikes a chord, a thousand listeners across America simultaneously lift their phones to call and agree, saying: "It's an honor to speak to you, sir."

In the pay of the corporations, at the service of the rich, he tells his people: "Nature is infinitely forgiving."

And the faithful listening on 600 plus radio stations nod in agreement, clap & whistle.

"It is impossible," wrote Mr. Menken, "to underestimate the intelligence of the American people."

xv.

Is the future with the Fat Man and his friends in the big cities ... the grain merchants, the oil men, the drug company execs, the insurance CEOs, the speculators, the investment bankers, all the ones buying politicians and writing laws? Will the day ever come when Mike Fink will return, ten yards tall as he was in his days on the river, and step into the Washington hog house that smells of death and cigar smoke, scan the human refuse and announce: "Holler, you varmints! Holler fer the Snapping Turkle! or I'll jump right down yer throats! I ain't had a fight in two days and I'm spilen fer exercise!"

Creekography

Ethan Hirsh

To get to know a creek really well takes the same effort as getting to know another person. Visit frequently, at all times of day, in all seasons and conditions and moods. Ask leading questions. Admire its change of dress. Be respectful when it's stirred up. Get to know its other acquaintances. Introduce your family. Provide support where you can. Share your secrets. Spend the night. And always, be a good listener.

Being a photographer of creeks and streams requires most of those steps, and then some. The old railroad crossing sign sums it up simply enough: "Stop! Look! Listen!" You have to do all three to take in the countless small details that deserve a photographer's time—not all of which will hold still.

The "Stop!" is important. No matter why you're on or in the creek, always take a break to really take in your surroundings. Breathe the water-charged air, inhale the smells. Set your brain on peak alert. It's also wise to plan each step carefully, and think what you'll do to keep your camera dry if (when) you lose your footing.

Be on the lookout for little visual cues. Stripped twigs in the water mean beaver. Crawdad shells say coon. Overturned moss and uprooted grass show force of the most recent flood. For fun imagery, there's always the interplay of light and moving water. Rapids and ripples. Drips and dribbles. Sparkles and splash.

Sounds can lead you to other visitors. Deer hooves on wet rocks. Ducks bathing beyond a curtain of mist. Captain Kingfisher on his dive-bombing run. A great blue coming in for a landing. An armadillo wading across to the other side.

On some shots you can take all day. Others you have to shoot really really fast. Whatever your luck, shoot often. Better to have too many to choose from than none that are quite right.

And before you leave, start planning your next date. You've just barely begun to be intimate.

1. The annual cycle of flowering trees, shrubs and wildflowers along Midwest stream banks lasts fully half the year. Dogwoods are among the first to put on their show.

2. In secluded streamways, roles can reverse late in the day. Sycamores look real in reflection, water becomes sky and light shines up.

3. Even the smallest streams can make white water after a heavy rain. Many waterways lie empty much of the year, coming quickly to life as land around them sheds rainfall.

4. Most fallen trees sooner or later float downstream during a flood. A few lodge permanently in mud and gravel and stabilize the streambed.

5. Pausing on the way to its next dining spot, a northern water snake takes a look around. Its diet consists almost entirely of small fish.

6. Not too small to cast shadows, young fish dart about in just a few inches of water. Their color blends perfectly with the surroundings.

7. Stretching ahead like a small rural highway, this creek in central Missouri invites the inquisitive to find out what's around the bend.

8. The Maine lobster's diminutive cousin, the crawdad, is readily found in midwestern creeks and streams, scooting backwards across gravel and hiding under rocks.

9. Well past surrendering its authority, the warning marker from a state-maintained low-water bridge rests among the rocks several feet under.

WATER
GAUGE
3
FT.
2
FT.

10. After a heavy morning mist has melted away, branches high above the creek put their drips into play on the water's surface.

11. Potholes form as gravel scours out cavities in a limestone streambed. The holes grow larger over time, providing hiding places for small creek dwellers.

12. In gravelly areas, creeks can pull a disappearing act when the water table drops. They go underground for a distance, then reappear in their usual channel.

13. Startling discoveries often await when you go creekwalking. Lodged in the rocks underwater, this bleached, leathery carcass was once a coyote.

14. The presence of a water strider translates into bold patterns of shadow and light as the insect speed skates from spot to spot.

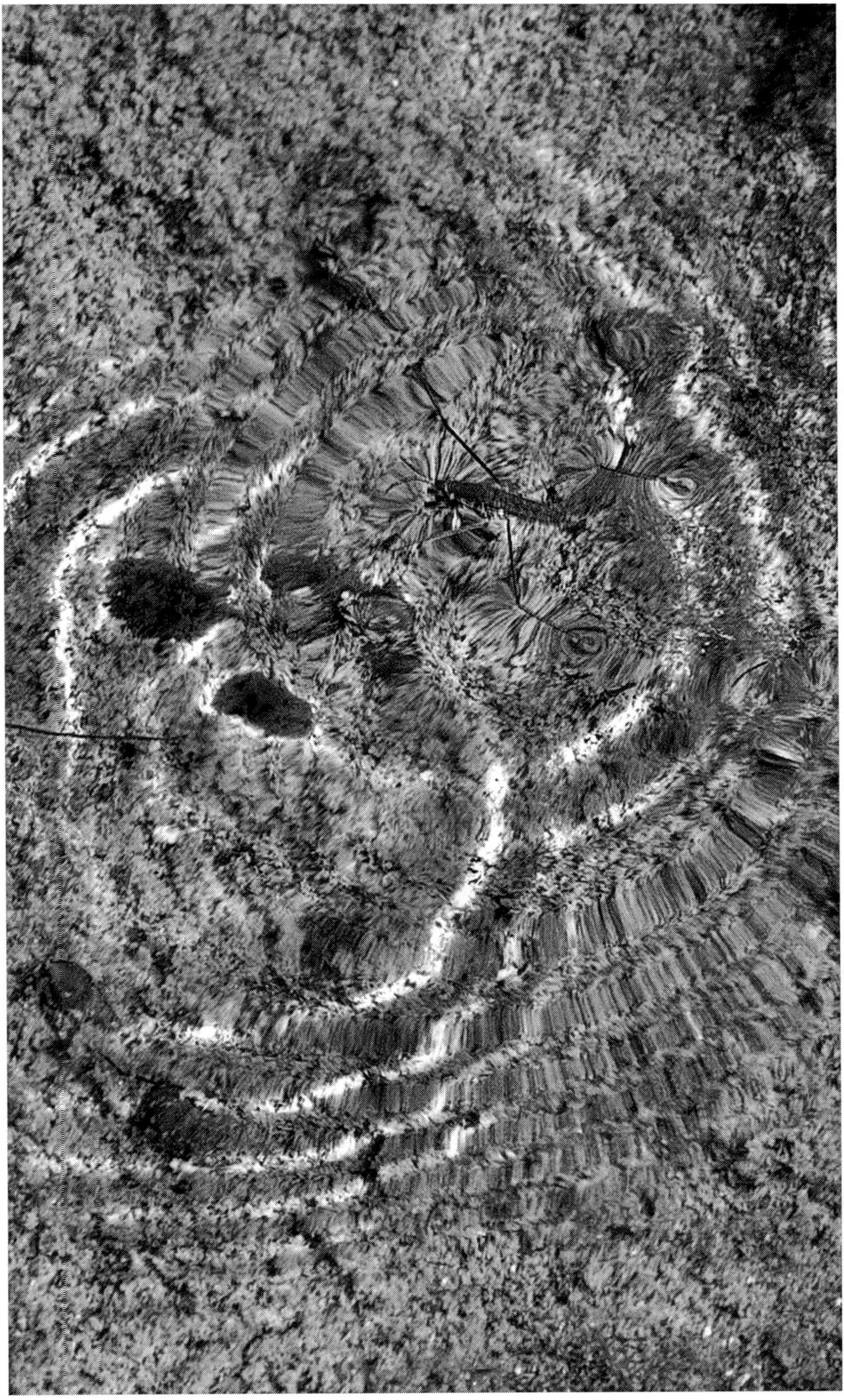

15. Unable to free its wings from the water's grasp, a butterfly makes waves during its last futile struggle to take flight.

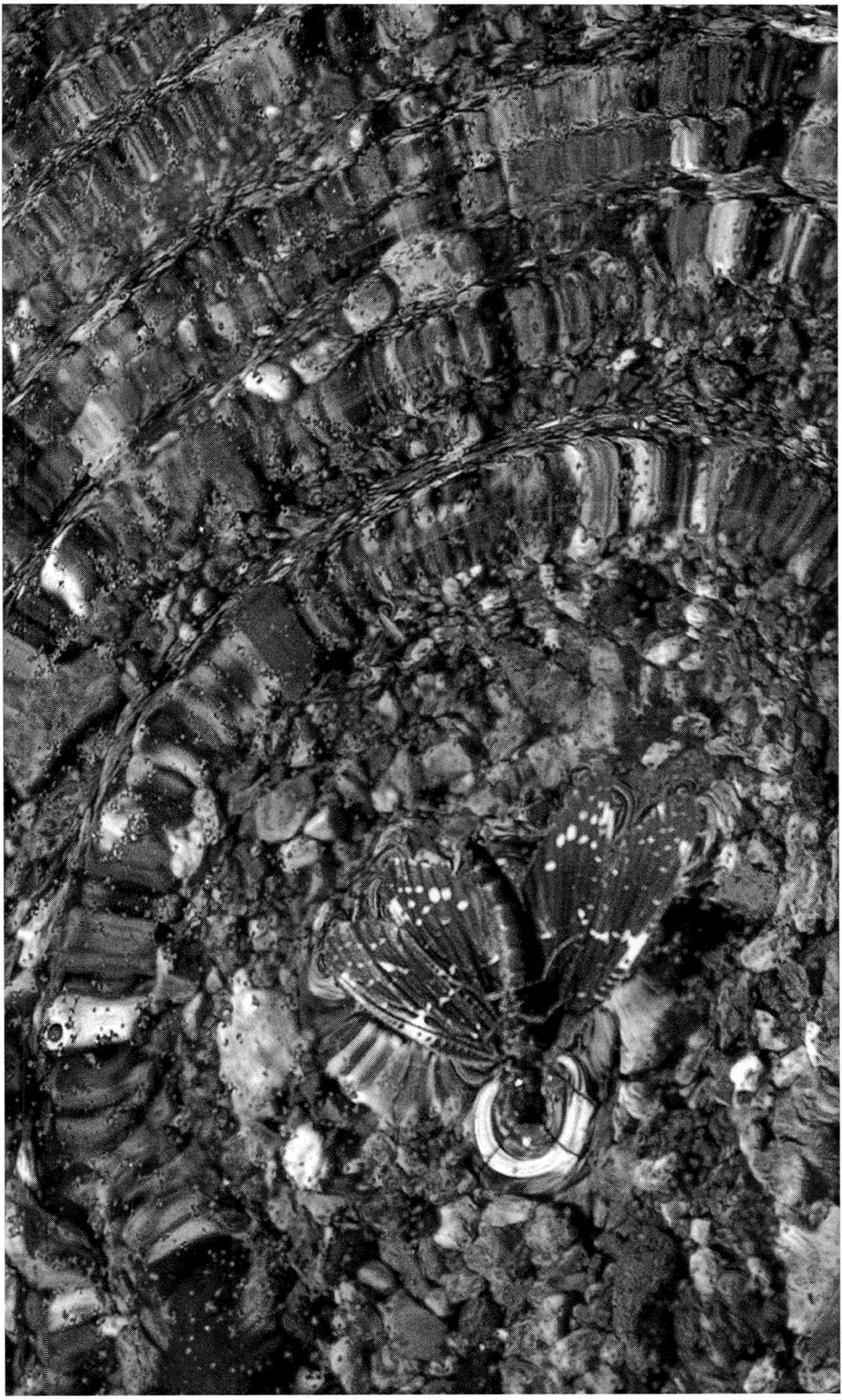

RIVERS

Gary Holthaus

I've known rivers, ancient dusky rivers.
My soul has grown deep, like the rivers.
—Langston Hughes

Inside each of us a river runs, perhaps more than one. For me there are at least two. One is small: Elk Creek flows indelibly through my childhood in northeast Iowa among hills too steep to plough or pasture. There were trees on the hills, fed by the creek: walnut, oak, hickory, occasional ironwood, linden and maple; there were raspberries, red and black, and currants. In summer, watercress choked the smaller tributaries, and in the fall, sumac flared brilliant in the undergrowth. Elk Creek ran clear and cool, and, unusual in the midwest, had rainbow trout that could be caught on a fly rod using single salmon eggs. Among my earliest memories are images of picnics, all of us in the old Ford driving down the dirt road past the old Spring Creek Mill built in 1867, its huge wooden wheel still turning in the creek in my childhood. I remember once riding on the wagon's high seat with my grandfather as we took the team and wagon down to the mill to get yellow corn ground into meal. He took me upstairs to see the great hand-hewn limestone wheels, turning almost imperceptibly. Along Elk Creek, near a wooden bridge, we would spread blankets under the trees to eat. It was cool under the great trees, even in August when the Iowa nights were so hot we could hear the corn grow, and there were squirrels to hunt in October when the leaves were aflame with approaching death. I still believe that October days hunting squirrels in the hardwoods along Elk Creek are among the finest days of my life.

The other river was huge. The Mississippi flowed below the bluffs at Dubuque, Iowa, where I was raised, through Lock and Dam #11. A few miles above Dubuque, the great river stretched nearly a mile across the widest point it would carve on its way to the Gulf. Ostensibly for flood control, the dams flooded thousands of acres that had never been flooded before; we called it "the best dammed river in the world." That river still runs somewhere in me. The first time I went duck hunting with my father, about age ten, we stood on the sloughy banks of the Mississippi in willows higher than my head, holding shotguns. There were plenty of ducks, thousands of them along this stretch of Mississippi flyway, but the day was too nice, the ducks flying too high. Occasionally a single duck would cross our line of vision, in range, and I would watch, mesmerized, as it sailed past. They appeared so suddenly, so unpredictably, and passed so quickly, it did not occur to me to shoot.

Once I thought this great river might take my life. The old German men of Guttenberg would come down with folding canvas camp stools and fish from the dock for carp, using worms, doughballs, or cheese for bait. They put their catch on a stringer and dropped the fish back in the water until it was time to go. The carp they had caught would swim around in the shadows alongside the pilings, like puppies on a leash. At the end of the day, the men would unstring their fish, gut them, throw them in a gunny sack and head home. Burton Jones and I pushed off below them one morning in a heavy, decrepit wooden rowboat. We were headed for game fish, bass, walleyes or northerns; carp were beneath our purist young dignity. The men murmured in German, and we knew they were talking about us; one pointed, others nodded in our direction with a knowing look, a rolling eye.

WALLEYES
Setting out to fish
I row in the sunlight
across the Mississippi.
Close against Wisconsin
a swifter current runs
and hauls us down against our will,
willows sweeping by,
boat swinging stern to.
Rowing the skin off my palms

I ship an oar out to look at a hand,
double it into a fist
and press it between my knees,
unable to believe the power
of this race that holds us,
and then, believing, row more;
remember the locks and dam below,
their silver rollers chewing the river
and row more, till — as if set free
by kidnappers, we slip out
into the ship's channel
but far down,
then have to pull again
against the whole Mississippi
until we are back on the Iowa side
without ever wetting a line
or catching a single wide-eyed pike,
emptier handed than I've ever been
in my whole life,
huge moon rising from Wisconsin
lighting the current behind us
as we haul up and over
the wet planks
and flap ourselves down
like unstrung carp
on the dock
at Guttenberg.

There may have been a third river. If I sound tentative about that, it is only because the third is not really a river, but a flow of water along the curb on the street in front of our house. When I was very young, dark thunder and lightning storms would besiege the seven hills of Dubuque, and we would go out in the street to play. The downpour would flood along the curb, the late afternoon sky to the west would flare with sheet lightning, and we would dam the river headed down to the iron gutters at the bottom of Bennett Street. Sometimes we carried rocks and gravel from the alley beside my house, but the easiest way was to lie down on one's side in the street, curl into a near fetal position with knees against the curb, body funneling the water against the tops of thighs to pool between thigh and curb and belly and chest. The rain was so warm on those summer days that it could lull one to sleep, right there in the street, thunder cracking as if in a dream. Latent in me still may reside an impulse, as in Hugo's great line, "to alter the river."

When I was 17, I used to walk, often at night and lonely, along a little stream near Hopkinton, Iowa. In winter, on moonlit nights, the bare black branches of the trees moved suggestively in the slightest wind, making patterns against the moon, their shadows shifting on the silver water or on patches of snow along the bank. The stream then seemed full of mystery and solace, a pull so powerful, alive and poignant that the river was wildly romantic, achingly erotic. Only women could have been more mysterious, elemental, and attractive. I wrote poems back then that would begin, "The river is a woman/dressed in black lace." I had never seen a woman in black lace.

Once, standing in a riffle, fly fishing on the South Fork of the Shoshone in Wyoming, I dropped a Joe's hopper on the quiet water just above. As I watched, intent, a fish the size of a cottonwood trunk rose to the surface, and the fly disappeared. I set the hook, but the fish did not run, did not need to; it could do whatever it pleased. What pleased it for a time was to settle back where it had been before. It did not seem to know it had my fly in its lip. I could not move it. I pulled carefully, to my right and again to my left; I waited impatiently, then bumped the bottom of the rod with the heel of one hand. I had no net, was up against a high cut bank; I looked around for a spot to do something with this creature from the river, but there was no place to land a fish till far downstream.

After a while I did not know whether to be frantic or bored, for the fish just sat there. I yanked on the line, as I might if I'd had a snag. I was tied to that fish, and the fish had a mind of its own, just sitting out there in the river, perhaps, by damn, not even knowing he'd been caught. How can you catch something that doesn't even know it's been caught?

I moved cautiously on the slippery rocks, making my way toward the fish, trying not to get water over the top of my boots. The fish, as if oblivious to the commotion I was making in the river, still did not move. Carefully, I reached down and pulled a rock from the river bottom. I threw it just beyond the place where my line entered the water; there was no response. I got close enough to see the dark form in the water. Perhaps it was not as long as my leg, but it looked as heavy. I had never been connected to such a fish before. Finally, as if making a judgement after long deliberation, perhaps sensing my presence or tiring of the pull on his lip or in response to a signal from a friend upstream, he began to move lazily against the current. I followed as best I could, keeping the line taut. He turned back, and I grabbed frantically for line; then suddenly he began to run wildly across the river. He caught me unprepared, line tore from the spool, but we stayed together, bound by the river and the line while he slashed around.

At last he began to drift back. I thought he was tired, but he was still vertical in the water, keeping his head upstream. He came down close to my left leg, just out of reach. I could see his brightness, then, through the clear water, the fins working easily. Holding the rod high, trying vainly to horse him just a little closer, I reached down to slip my left hand into his gills. Perhaps he saw the shadow; perhaps I pulled too hard. With a simple flip of his head, he broke the leader and swam leisurely away, apparently not the least bit excited. I stumbled back, almost falling, legs shaking, hands trembling and moved downstream to where I could finally get out and sit on the bank. What if that trout had decided to attack; how would I have defended myself? I was not sure they made a gun of sufficient size to stop him. There are things in the river beyond our human control.

The old poet might have said, "To those who in the love of rivers, hold communion with their visible forms, they speak a various language." He did not, but anyone who has known rivers knows they do speak, that they offer many levels of discourse. The river, as Norman Maclean says in *A River Runs Through It*, carries on a continual conversation with itself. Under the rocks of the river are the words, The Word—in the fly-fishing, Johanine sense—the word that lies at the bottom of all creation. My rivers may be less theological, more personal, but the voice of Elk Creek, and the trees along its banks, have not diminished with age and distance.

I have no way to know what river or rivers flow in others, but I suspect there is at least one. Your rivers and my rivers together become the river. Not a faceless generic river, but The River: Tigris, Euphrates, Platte, Mississippi, Big Horn, Nile, Niobrara, Tongue, the Yellow, the Missouri, the Yellowstone. The river is embedded in our soul not only because it has slaked our thirst and given life to the grain and game upon which we depend, but because our own human blood has been drained back into the river. We have absorbed rivers, carry them in our living tissue, and they have absorbed us, carry us in their spiraling currents and eddies.

The blood of our ancestors, our kin, has seeped into the Rosebud, the Big Hole, the Ganges; into Sand Creek, the Amazon, the Orinoco, and their names ring in our minds and stir us still. "I've known rivers..." wrote Langston Hughes. "My soul has grown deep, like the rivers."

Perhaps Plato was right: there is an ideal river that is at once the embodiment and realization of all our rivers. Riparian Platonism is not a branch of philosophy I've ever heard of, but perhaps there ought to be one, for we are talking about that from which we come and which sustains us while we are here. It seeps into the aquifer of our souls and is the source of our very being.

Thinking again of those earlier days, those rivers, questions come to mind: Do we have an interior life that parallels our exterior life in the world? If so, will altering the river also alter the soul, the mind, the psyche, the way we see the world? If the river really does run deep in us, what happens if we block it? Will that somehow dam our souls as well? The poet said that nature "speaks a various language." If we shut down the natural discourse of the river with the land that borders it, do we thus end our own ability to hold discourse with the natural world around us? If we kill the spirit of the river, will our own spirits die? Do our souls need moistness, just as our bodies need moisture? Like the river, we are mostly water, body and soul, and the land still marks the limits of each.

From the beginning we have been among the creatures who settle along the streams and lakes or on the nearby terraces above them. For a million years, perhaps even 3,000,000, we have not settled, or even stopped for the night, far from water; our heritage is one of shore-dwelling. Only in recent years have we been able to distance ourselves somewhat, been skilled enough to bring water to locations far removed from the river in quantities sufficient to sustain large communities. We know the Babylonians did it, and their irrigated gardens are said to have been one of the Seven Wonders of the Ancient World. Unfortunately, they couldn't keep their ditches from silting up, and they could not sustain themselves. The Romans may have been the best at diverting and transporting water, and we can still see parts of their great viaducts standing above the narrow valleys in central Italy and southern France. But they could not control their politics and were not able to sustain themselves.

It would seem that therein lie two clues to understanding what we must do to maintain our own culture: better understand the implications of our technology, and discipline ourselves.

In our own country, as we rather timidly left our forested East to move out onto the arid western savannahs, rivers came to have a dual significance: we used them then not only to sustain ourselves, but as navigation aids. We followed the rivers, such as the Missouri, in part because they were going our way (coming our way), and in part because we couldn't leave them with any assurance there would be water elsewhere. The sun was so persistent, the ground so dry, the storms so brief and fierce that water ran off or soaked in before we could think to contain it. When we did stop and settle, we settled along a river.

Somewhere in almost every Western town that has survived there is a river. Sometimes it is hard to find: Denver's Cherry Creek is no longer the main feature of Cherry Creek, surpassed for years now by small stores and the infamous Tattered Cover Book Store. But sometimes there is a small parkway along the banks, or a necessary bridge. Frequently, as in Santa Fe, the river is dry, the bottom concrete, the refuse and discards of town more apparent than they should be. Most often the stream that meant life to the early community is now considered unimportant and insufficient, and the water that keeps the town growing comes from more distant places; most residents can't tell you where their water comes from.

Water or the lack of it, rivers and their absence are among the defining characteristics of the West. Folks as knowledgeable and experienced as John Wesley Powell and Wallace Stegner have urged us to think of the West as arid land, and they were right to do so. What we know from our Western past and our own present experience is that water is one of the critical resources, that in the West it tends to be scarce, and that it has been and is the focus for intense argument and debate, sometimes the cause of violent confrontation and death. Thousands of homesteaders were defeated by the failure of rain to come, even though they believed what they had been told: that rain follows the plough, that rain is tied to settlement, and, as the West fills up with people, the rain will automatically increase. Surely such faith in rain, or such willingness to believe our fellow human hucksters, gives us ample reason to ask if there is intelligent life on Earth.

Not all men believed that booster story about water. Fearing that they would not have enough, some who had early rights to the river took it all and left none for their neighbors downstream. Mary Austin has a story about a man who sat on his headgate with a gun and warned off his neighbors who came to see what had happened to the water. When one downstream owner came armed, the man shot him five times in the chest. The next man to own the neighbor's plot was smarter. He sent his wife to the headgate in the middle of the night. She was a formidable woman of considerable size, and when the farmer came up in the morning to protect his water, she was sitting on the headgate knitting, a long-handled shovel across her lap, the water already turned into her husband's fields. The man, true to the code of the West, would not shoot a woman, and she looked so strong and armed with a shovel, he would not attack her one-on-one either.

Whatever we know or think we know about water, however much we may involve ourselves in studying water issues, we can be sure that the final word has not yet been said about it. In Utah a year or so ago, I heard about two retired ranchers who were walking together. They huffed to the top of a dike and found a shallow pond. One man sat down to rest; the other walked all around the pond and came back to report, "You know Howard, there's about a foot and a half of water in the bottom of that pond." "Yes," said Howard, "That's just where it would be all right."

We remember our rivers, even if we do not honor them. They enliven our sense of the past, inform our sense of the present. Driving west on I-80 across Wyoming, any Westerner is reminded that south of the highway about 40 miles, down on Henry's Fork, is the site of the first free trappers' rendezvous, held in 1825. One also remembers that a little north, on the Green River, is the site of one of the last, in 1835. Such a short span to have such a mythic hold on the American imagination. Perhaps the strength of that memory's grip lies in the idea that nearly every historic incident held in our Western recollection of those days is connected to a river. Across Black's Fork, just to the south of the divided highway, one can see the location, such a good site, of Jim Bridger's old fort and trading post, the valley, even in high summer, still a mild riparian green in the midst of the desert gray and beige.

But memory is insufficient to keep the river moving. If one continues along I-80, swinging down into Utah, the land drops off into Echo Reservoir. Late summer the reservoir is so low, the east side is a long slope to the water, the lower end almost barren except for the trickle that drains into Echo River. These are apt names; both reservoir and river are but echoes of past water. What lures a fisherman there now? The memory of past fish? The men standing on either side of the Ford Bronco parked in the willows are so far away one cannot actually see what they are doing, but memory again comes into play and lights our knowledge of what is happening. By their intent posture, one knows that these men have fishing poles leaning inside their arms. They have just threaded line through the guides, are knotting leader or monofilament. One needn't keep watching to see in the mind exactly how the hands tilt upward from the wrist, how the fingers move; to know that in a few seconds a hand will fall from chest height, the forearm then rise again to grasp the invisible pole, holding it next to a thigh, almost horizontal to the ground, tip raised slightly. The left leg will lift to step toward the path through the soft willows to the headwaters of Echo River. No need to practice prophecy to see the future here; it is revealed by memory of our own moves. For the rest, whether they will catch fish or not, one can be as fancifully imaginative as any novelist or historian. One of those men, I think, will hook a fish so big it will drag him down the creek, through the canyon, into the Great Salt Lake. Tomorrow searchers will find only his hat, floating toward Antelope Island, and he will dwell in the house of the sea forever.

If the river stirs our fantasy, it is neither memory nor fancy, but reality, that tells us the river is our parent. It was here before we were; in its soft lilt we were born. The river nourished us when we were new on the earth and sustained us as we grew. But like all parents, the river is growing old as we grow older. Now, as sometimes happens with our own aging progenitors, the roles are reversed, and the life of the river is in our hands. We must care for it, as once it cared for us.

Though water has been one of the most complex and divisive issues in the West, and is about to become so across all the nations, it is also what draws us together and allows us to know our place. Water is the common bond that we share with one another and with all other creatures. It is one of the gifts that we have to have in order to live and that therefore we must share. Historically, if there has been any healing in the world, it has come from the rivers. It is "by the rivers of Babylon that we lay ourselves down, by the cooling waters we lay down." Water is not only a difficult and divisive issue; it is a source of our unity, a symbol of rest and health for mind and body, and our great hope.

Contributors

DAVID HAMILTON edits *The Iowa Review* and teaches English literature at the University of Iowa. With degrees from Amherst College (AB) and the University of Virginia (PhD), he taught in Colombia and at the University of Michigan before taking his present position. The University of Missouri Press published his *Deep River: A Memoir of a Missouri Farm* (2001).

JOHN T PRICE is a writer of creative non-fiction and holds the PhD in English and the MFA in Nonfiction Writing from the University of Iowa. He is the author of *Not Just Any Land: A Personal and Literary Journey Into the American Grasslands (Univserity of Nebraska Press).* His personal essays have appeared in *Creative Nonfiction, Orion, In Brief, Organization and Environment, North Dakota Quarterly, Echoes, The Florida Review, The Christian Science Monitor,* and *Best American Spiritual Writing 2000* (Ed Phillip Zaleski, Harper, 2000). He was awarded a 2004-2005 fellowship by the National Endowment for the Arts and was nominated in 1998 for the Pushcart Prize. An Associate Professor teaching English and Creative Writing at the University of Nebraska-Omaha, he lives in Council Bluffs, Iowa.

LISA KNOPP grew up in Burlington, Iowa. She is the author of *Flight Dreams: A Life in the Midwestern Landscape* (Univ. of Iowa Press), *Field of Vision* (Univ. of Iowa Press), and, most recently, *The Nature of Home* (Univ. of Nebraska Press). Her essays have appeared in many places, including *Shenandoah, Michigan Quarterly Review* and *ISLE* (Interdisciplinary Studies in Literature and the Environment).

ROBERT WOLF, Free River Press director, is the author and editor of numerous books, including *An American Mosaic: Prose and Poetry by Everyday Folk* and *Jump Start: How to Write from Everyday Life,* both published by Oxford University Press, and *The Triumph of Technique: The Industrialization of Agriculture and the Destruction of Rural America.* He lives in Lansing, Iowa.

Environmental photographer **ETHAN HIRSH** works to capture on film the paintings created by nature such as the exotic inner detail of flowers, the physical intricacy of insects, the life cycles of flora and fauna, and the interaction of light, land and water. Ethan's goal is to enable viewers to see something they haven't seen before, including the often abstract beauty that comes and goes in the scenery around us. After a three decade career in corporate communications, Hirsh began a serious effort to capture the flora and fauna of the Missouri Ozarks on film.

GARY HOLTHAUS received an Individual Fellowship for Poetry from the National Endowment for the Arts in 1990. He has been a commercial fisherman in Alaska, a big-game guide in Montana, a wheat packer for Quaker Oats, a school teacher and worked "too long," he says, "moving steel beams around for Iowa Steel and Iron Works." His books include, *An Archaeology of Home* poetry about memory, family, and growing up on an Iowa farm. As well as *Wide Skies: Finding A Home In The West* (University of Arizona Press).

The Harvest Lecture is now in it's ninth year and run primarily by the *Standing By Words Center* which is a nonprofit organization whose mission is to promote creative and explorative thinking on how people can best live together in the community they inhabit. This purpose rests on the recognition that there are tensions between human, social and economic systems as well as between the variety of natural and ecological systems of the earth. We work to fulfill our mission by integrating the perspectives of the arts, humanities and sciences in specific programs, publications and events that focus primarily on Iowa, but also the wider context of the Midwest. To find out more please e-mail to sbwc@standingbywords.org or write to: *Standing By Words Center* 205 N. Front St. North Liberty, Iowa 52317

For more information on past and future events please check out our web site at www.standingbywords.org

Additional Harvest Books

Prairie Weather

1-888160-17-9, $10

At the crossroads of the elements, just above our heads whirl wild and other-worldly tornadoes, the wailing of the wind and bone drying droughts, all of which inspire our senses and idea of place. In our region of four seasons we can learn much from our weather. Writing and photographs by Jim Heynen, Mary Swander, Deb Marquart, Amy Kolen, Ron Sandvik, Mark Petrick, Ethan Hirsh, Robert Sayre, Tom Dean, Patrick Irelan, Michael Harket, Scott Cawelti and a foreword by Denny Frary.

Living With Topsoil

1-888160-99-3, $9.95

A full fledged exploration via Iowa's finest authors into living with our world famous fertile topsoil. New and valuable writing by Mary Swander, Connie Mutel, Michael Carey, Patrick Irelan, Thomas Dean, Larry Stone and Tim Fay and an introduction by Steve Semken. Jose Ortega y Gassett once wrote, "tell me where you live and I'll tell you who you are." You'll find out what it means to live in the land of amazing topsoil once you read this book.

Finding The Center of the World

1-888160-05-5, $10.95

Featuring a timely essay entitled 'Patriotism and Place., by Peter Bakken of the Au Sable Institute. Mr. Bakken draws inspiration from two relevant sources: the spirit of Aldo Leopold's land ethics and Joseph Sitler, well known for his theological concerns on the care of the earth. Also included in this charming book are poems on place by Univesity of Iowa Hospital Chaplain Nancy Adams-Cogan and photographs by Rod Strampe. Finding the center of a place is akin to creating the world.

The Good Earth: Poets of the Prairie

1-888160-09-8, $9.95

Surprisingly, there is a very intriquing tradition of prairie poetry here in Iowa. Featuring the prairie based works of legendary poets Paul Engle, James Hearst and William Stafford. Examined, respectively, by scholars Robert Dana, Denise Low and Scott Cawelti with a foreword by Iowa farmer/poet Michael Carey. In the tradition of placed-based stories this book finds connections between spirit and place. As if this isn't amazing enough, this collections also includes one previously unpublished poem by Iowa's famed, writer's workshop director Paul Engle.

Prairie Roots: Call of the Wild

1-888160-12-8, $10.95

Explores the meanings of the wild in the Midwest and features perhaps the last "published" essay by the late Minnesota author, Paul Gruchow. As well, this intriquing collection explores other facets of Iowa and the prairie lanscape. A very fascinating examination of landscape art by Joni Kinsey, the "grid" system laid upon our land with Robert Sayre as well as poetry by Mary Swander, hearing the call of geese and the value of home by Thomas Dean not to mention a strange discovery of giant worms by Steve Semken. Photography by Rev. Howard Vrankin.

Natural Prayers: The Tin Prayer

1-888160-01-2, $11.95

An exploration into the meaning of place and the need to care for where we live and come from. In many ways this book is a regionalist manifesto. Written by Steve Semken while writer in residence at the Island Institute in Sitka, Alaska with grants from the Iowa Arts Council.

Ordering Information:
Books can be ordered directly from our web site at
www.icecubepress.com
or by mail (check/money order) by sending to
Ice Cube Press
205 N Front St.
North Liberty, Iowa 52317-9302

(shipping $1.50 first book, .25¢ each additional)

The Ice Cube Press began publishing in 1993 to focus on how to best live with the natural world. Since this time we've been recognized by a number of well known writers including Gary Snyder, Gene Logsdon, Wes Jackson, Annie Dillard & Barry Lopez. We've published a fair number of well known authors as well, including, Mary Swander, Jim Heynen, Stephanie Mills, Bill McKibben & Paul Gruchow. Check out our books on our web site, with booksellers, or at museum shops and discover why we are dedicated to "hearing the other side."

Ice Cube Press 205 N Front Street
North Liberty, Iowa 52317-9302
p 319/626-2055 f 413/451-0223
steve@icecubepress.com
www.icecubepress.com